A Problems Approach to
Introductory Biology

A Problems Approach to
Introductory Biology

Brian White
Department of Biology
University of Massachusetts Boston
Boston, Massachusetts

Michelle Mischke
Department of Biology
Massachusetts Institute of Technology
Cambridge, Massachusetts

ASM
PRESS

Washington, DC

Copyright © 2006   ASM Press
American Society for Microbiology
1752 N St., N.W.
Washington, DC 20036-2904

Library of Congress Cataloging-in-Publication Data

White, Brian T., 1963–
  A problems approach to introductory biology / Brian T. White, Michelle D. Mischke.
     p. cm.
  ISBN-13: 978-1-55581-372-7
  ISBN-10: 1-55581-372-0
  1. Biology—Problems, exercises, etc. I. Mischke, Michelle D. II. Title.

QH316.W45 2006
570.76—dc22

                              2006008192

10 9 8 7 6 5 4 3 2 1

Address editorial correspondence to ASM Press, 1752 N St., N.W., Washington, DC 20036-2904, U.S.A.

Send orders to: ASM Press, P.O. Box 605, Herndon, VA 20172, U.S.A.
Phone: 800-546-2416; 703-661-1593
Fax: 703-661-1501
E-mail: books@asmusa.org
Online: http://estore.asm.org

Table of Contents

Solutions to Problems can be found on the enclosed CD-ROM.

Introduction

Nearly anyone with an interest in dance can recognize the tango. Being able to recognize the tango is only the beginning. The difference between recognizing the tango and dancing the tango requires learning the fundamental steps and practicing these steps until they become fluid. This is also true in biology; recognizing a biological concept is just the beginning. Like dancing, fully understanding a biological concept requires knowing the fundamental steps and practicing these steps until they become automatic. Finally, actually dancing the tango is the only way to discover which steps you know well and which need more practice.

The problems in this workbook ask you to practice the fundamental steps in biology. In addition, they are designed to help you measure your learning and target your studies to those areas that need the most work.

We have created problems for three major topic areas: genetics, biochemistry, and molecular biology. The study of genetics is well suited to problem solving, and we begin the workbook by reinforcing this natural relationship. In the subsequent chapters, basic concepts of biochemistry and molecular biology are explored in analogous ways. Note that the chapters are independent and can be approached in any order.

Within each chapter, we present three types of problems: problems based on actual data or situations, problems based on a simplified version of actual data, and problems based on hypothetical or fictitious data. Some are pencil-and-paper problems, while others use computer software to simulate or display biological concepts. The intentional redundancy in this workbook provides many different views of each concept; in our experience, the combined views of a concept create a deeper understanding of the material.

Perhaps the most important feature of this workbook is that it represents a wonderful collaboration between the authors and the many students with whom we have worked. Each of these students provided insights, asked questions, uncovered misconceptions, and made mistakes; this information is the foundation for this book. This workbook is not so much a transfer of knowledge from author to student, but more a gift from one generation of biology students to another.

How To Use This Book

This book is designed to support a college level introductory biology class or an advanced high school biology class. For each topic area, we expect that the concepts will be introduced in class and elaborated on in the textbook. Once this foundation is in place, working these problems will solidify your knowledge and give you feedback on your grasp of the concepts in each topic area.

Your first step should be to attempt the diagnostic problem at the start of the appropriate section of the book and compare your method and answer to what is given in the workbook. If the diagnostic question goes smoothly, proceed to the additional questions in the chapter. If you struggle with the diagnostic question, review your textbook or consult the teaching support for your class.

The remaining problems will be most valuable if you allow yourself to make mistakes. You must write out your best answer before looking at the solutions. Once your best answer is drafted, check the solutions. The solutions can be found in the "Solutions To Problems" folder on the CD-ROM. The solutions guide you through the problem from beginning to end, and your tendency will be to read the entire solution. Don't! If the solutions uncover a misstep in your answer, stop reading. Go back to the problem at this step and try again. Relying on the solutions defeats the purpose of the problems. Remember... just because you can recognize the tango doesn't mean you can dance it! Use the solutions to reinforce the correct steps and modify the faulty ones. If you find that you must look at the solutions to complete several problems in any section of the workbook, then further preparation is needed. Tedious as the method outlined may seem, these problems and this approach will build the background that allows you to move fluidly through difficult biological concepts.

Using the Software on the CD-ROM

In General

All the software programs on the CD are supplied in both Mac OS X-compatible and Windows-compatible forms. You should use the software in either the "Windows" or "Mac OS X" folder as appropriate. You can run the programs off of the CD or copy the contents of the appropriate folder to your hard drive.

Minimum System Requirements

The software has been tested and found to run acceptably fast on the following systems. It will certainly run faster on more capable systems, and it *may* run on less capable systems.

Mac OS X	**Windows**
OS 10.3 or higher	Windows 98 or higher
233 MHz processor	400 MHz processor
384 MB RAM	192 MB RAM
50 MB free hard drive space	50 MB free hard drive space

Software Updates

Updated versions of the software can be found at:
        http://intro.bio.umb.edu/APAIB/
You can check there to see if any newer versions have become available.

Java

The software on the CD requires the Java Computer Language. Java comes built in as part of Mac OS X; not all versions of Microsoft Windows have Java built in. Java is not available for Mac OS 9 and lower. Java for Windows is available for free. These are instructions for setting up Java on a Windows PC.

See the next page for more help in installing and using Java.

1) How can I tell if my PC has Java installed?
Double-click on "CheckYourComputerFirst" in the "Windows" folder of the CD-ROM.
You will see one of three things:

- An "Executor Error" window like this:

  This means that Java is not installed on your computer. You need to install Java
  as described in (2) below.

- A <u>green</u> "CheckYourComputerFirst" window like this:

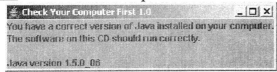

  This means that you have the correct version of Java on your computer. The
  software on the CD-ROM should run correctly.

- A <u>red</u> "CheckYourComputerFirst" window warning you that you have an
  improper version of Java installed. You need to install Java as described in (2)
  below.

2) How do I install Java on my PC? You will only have to do this once for any particular
PC.
  1. Look in the "Windows" folder on the CD.
  2. Double-click on the icon marked "jre-1_5_0_03-windows-i586-p.exe" and
     follow the directions.
  3. You may need to reboot; do so.
  4. The software should run just fine now.

You can also download for free the very latest Java for Windows and other platforms
from http://java.sun.com/getjava.

Acknowledgments

We gratefully acknowledge the help of many people; this book would not have been possible without them:

- all the many Teaching Assistants at the Massachusetts Institute of Technology (MIT) and the University of Massachusetts, Boston (UMB), who helped make these problems become real and valuable.
- the software team who wrote the Virtual Genetics Lab software at UMB (Nikunj Koolar, Naing Naing Maw, Wei Ma, Chung Ying Yu, and Ethan Bolker).
- the people at UMB who patiently tested and reviewed the software (Eden Medaglia, Nicole Weber, and Molly Carrier).
- the people at MIT who patiently edited the pencil-and-paper problems (Megan Rokop and Aurora Burds Connor).

Chapter 1:

Genetics Problems

Genetics Problems

If you were a geneticist, you would study genes. This statement is particularly true today in an era in which, in some organisms, the entire genome has been sequenced and nearly every gene has been identified.

Today, once a gene has been identified, the protein (or RNA) that it encodes can be inferred. Yet even when the gene and the resulting protein are known, the <u>function</u> of that gene may remain a mystery.

If you were Gregor Mendel, you would have studied mutant plants. In 1866, Gregor Mendel was performing what many see as the first genetic experiments, and he did not even know genes existed! Mendel postulated the existence of "particles" (genes) whose function he could draw conclusions about. If his mutant had yellow seeds instead of green seeds, then a "particle," i.e., a gene, controlled seed color, and this gene was altered in his yellow mutant.

Modern day geneticists use a wide array of tools and techniques. They study genes by searching vast databases of genetic information, but they also study genes much as Mendel did, by beginning with a mutant that is clearly distinguishable from what is standard or normal. In this chapter, we have given you different genetics problems that will allow you to practice and build on the concepts introduced by Gregor Mendel. The very first question is meant to be diagnostic. Work through the problem on your own and then look at our approach to solving it. If any of the underlined terms are unfamiliar, please consult your book's chapter on Mendelian Genetics.

(1) PROBLEMS INVOLVING ONLY ONE GENE
In general, the problems in each section begin with the most straightforward and become more complex as you proceed. They are grouped hierarchically; that is, Problems 1.1.x all deal with one gene, two alleles, simple dominance, while Problems 1.2.x all deal with more complex problems.

Diagnostic Question:

You are given two blue beetles and two black beetles.

a) Cross 1: You <u>mate</u> (or <u>cross</u>) blue beetle #1 to black beetle #1 and obtain 220 black beetles in the F_1 <u>generation</u>.

- What is the <u>dominant</u> <u>phenotype</u>?

- What is the <u>recessive</u> <u>phenotype</u>?

- What are the <u>genotypes</u> of the two parents and the offspring? Be sure to indicate which allele is associated with the dominant phenotype.

b) Cross 2: You mate blue beetle #2 to black beetle #2 and obtain 55 blue beetles and 65 black beetles. What are the genotypes of the two parents and the offspring?

c) Cross 3: You mate two black offspring produced in Cross 2. What are the possible genotypes and phenotypes of the offspring? What are the expected proportions of these genotypes and phenotypes?

Answer to Diagnostic Question:

a) Cross 1: You mate blue beetle #1 to black beetle #1 and obtain 220 black beetles.

What is the dominant phenotype?
Because there are a large number of offspring, all of which are black, you can assume that black is the dominant phenotype.

What is the recessive phenotype?
Blue.

What are the genotypes of the two parents and the offspring? Be sure to indicate which allele is associated with the dominant phenotype.
Because black is the dominant phenotype and there are a large number of offspring, all of which are black, you can assume that the black parent is <u>homozygous</u> for the <u>allele</u> associated with the dominant <u>phenotype</u>. By convention you would use uppercase letters for the <u>allele</u> associated with the dominant phenotype. In this instance we have assigned the genotype BB to black beetle #1. The genotype of blue beetle #1 must be bb; otherwise, it would not have the blue phenotype.

b) Cross 2: You mate blue beetle #2 to black beetle #2 and obtain 55 blue beetles and 65 black beetles. What are the genotypes of the two parents and the offspring?
*Some of the offspring have the recessive blue phenotype and must therefore have the genotype bb. Thus, the genotype of blue beetle #2 must be **bb** and the genotype of black beetle #2 must be **Bb**. Also, based on (a), you know that blue has the genotype bb.*

c) Cross 3: You mate two black offspring produced in Cross 2. What are the possible genotypes and phenotypes of the offspring? What are the expected proportions of these genotypes and phenotypes?
The black offspring from Cross 2 must have the genotype Bb, so Cross 3 is Bb ✕ Bb.

	B	b
B	BB black	Bb black
b	Bb black	bb blue

The <u>genotypic ratio</u> is 1 (BB) : 2 (Bb) : 1(bb).
The <u>phenotypic ratio</u> is 3 black beetles to 1 blue beetle.

(1.1) One gene; two alleles; simple dominance

In this section, we deal with models that consider only a single trait or characteristic (phenotype), for example, green seeds or yellow seeds. The trait of interest is determined by a single gene that has two alleles.

Problems:

(1.1.1) Consider some hypothetical flowers in which color is controlled by one gene and green color is dominant to blue color. **G** is the symbol for the allele associated with the dominant phenotype, green flowers, and **g** is the symbol for the allele associated with the recessive phenotype, blue flowers.

Give the expected ratios of offspring from the following crosses:

a) GG × GG

b) gg × gg

c) Gg × gg

d) Gg × Gg

e) Green × Green (note that there may be several possibilities here; give them all)

f) Blue × Blue (note that there may be several possibilities here; give them all)

(1.1.2) For each of the following sets of data, give a genetic model that explains all the data. A genetic model contains the following:
- the number of genes and alleles involved, e.g., "tooth shape is controlled by one gene with two alleles."
- a statement of which phenotype is dominant and which is recessive.
- symbols denoting each allele such that uppercase letters are associated with the dominant phenotype and lowercase letters are associated with the recessive phenotype.
- the genotypes of all the individuals involved.

a)   Cross 1: Red-eyed mouse × white-eyed mouse

gives F_1:   all red-eyed

Cross 2: Red-eyed F_1 × red-eyed F_1

gives F_2:   36 red-eyed
13 white-eyed

b)   Cross 1: Long-eared mouse ✕ short-eared mouse

gives F₁: 12 long-eared
10 short-eared

Cross 2: Long-eared F₁ ✕ long-eared F₁

gives F₂: 34 long-eared
14 short-eared

(1.1.3) Achondroplasia is a form of dwarfism controlled by one gene with two alleles. Two achondroplastic dwarfs marry and have a dwarf child and later have a second child who is of normal size.

Based on this:

a) Is achondroplasia a recessive or a dominant phenotype?

b) What are the genotypes of the two parents?

(1.1.4) Assuming that the following traits involve only one gene with two alleles, give all models that are consistent with the data. For each model, indicate the genotypes of the individuals involved (indicate any ambiguities), which phenotype is dominant, and which phenotype is recessive.

a) Red fly × red fly gives one blue fly progeny.

b) Brown cow × white cow gives one brown cow progeny.

(V1) Virtual Genetics Lab I The Virtual Genetics Lab (VGL) is a computer simulation of Genetics in a hypothetical insect that allows us to perform virtual genetic experiments. It has a variety of features that we will introduce gradually. In this first problem, you will use VGL to generate appropriate offspring from a simulated cross of two individuals that you select. This simple use of VGL is designed as practice with the genetic models we have discussed and as a warm-up for later VGL problems.

     1) Find the VGL program on the CD-ROM in the "**Genetics**" folder.
     2) Launch the VGL program by double-clicking the **VGL** icon (it looks like a small winged insect). Note that the name VGL may be followed by numbers indicating the version, for example, "VGL1.3.jar"; this is the VGL program. The VGL program will launch and you will see a large blank screen with this menu bar:

     3) Because you will be working on a VGL problem that has already been started, choose "Open Work…" from the "File" menu. Select the file called "**Problem1.wrk**" from the list that appears and click "Open"; alternatively, you can double-click "**Problem1.wrk**" in the list. You will see this:

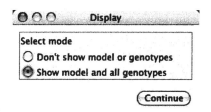

     4) Choose "Show model and all genotypes" and click "Continue." This is called "Practice Mode" because you can see the underlying genetic model and the genotype of any individual insect. In later problems, you will need to figure these out on your own.

You will see a cage containing some creatures:

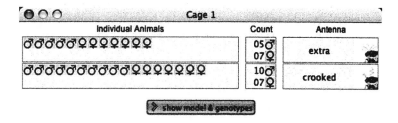

These creatures have a random assortment of genotypes. They represent a population of flies that might have been collected in the wild. They are not necessarily pure-breeding, nor is it clear if any are the offspring of any others.

5) Click on the button marked "> show model & genotypes" and you should see this:

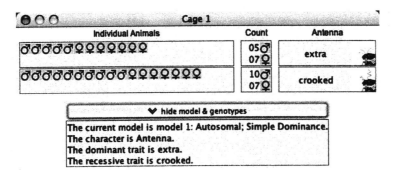

This gives the details of the genetic model for the inheritance of Antenna type in this particular problem. Each time you open Problem1.wrk, you will get this same genetic model, although the offspring of a particular cross may be different. In later VGL problems, the genetic model will be selected randomly.

The model can be written in the usual form:

Phenotype	Associated allele
extra antennae (dominant)	E
crooked antennae (recessive)	e

or:

Phenotype	Genotype
extra antennae	EE
extra antennae	Ee
crooked antennae	ee

You should do this as a matter of course for *every* VGL problem that you solve: every time you try out a new genetic model to explain your data, write out the model explicitly as shown above. This will help you to understand more clearly the different types of genetic models; it will make careless mistakes less likely.

In the cage, the insects are sorted by phenotype and sex. The left-hand column contains individual male and female symbols that represent individual insects that you can select for crossing. The next column shows a count of each phenotype and sex. In this case, there are:

     5 males with extra antennae
     7 females with extra antennae
   10 males with crooked antennae
    7 females with crooked antennae

The right-hand column shows a picture of each phenotype; you can click on it for a larger view.

You can find the genotype of any insect by putting the cursor over it and leaving it there for a few seconds. A box will pop up giving the details of that insect's genotype. For example, if the box shows:

Genotype: crooked; crooked

This means that the genotype is:

crooked – the first allele is crooked (e)

crooked – the second allele is also crooked (e)

therefore, this individual has genotype ee.

6) You can select an individual for crossing by clicking on it. Select one male and one female insect of either phenotype from any cage. Find their genotypes as described above. Predict the ratio of extra:crooked offspring using a Punnett square.

7) Cross the two parents you selected in part (6) by clicking the "CROSS" button in the VGL toolbar. A new cage should appear containing the offspring of your cross. For example, the cage below was created by crossing an Ee (extra) male with an ee (crooked) female; our model would predict 50% Ee (extra) and 50% ee (crooked) offspring:

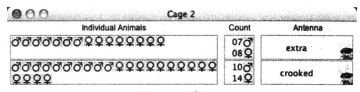

Parent ♂ (1) extra ♀ (1) crooked

The cage in the figure shows:

15 (7 male + 8 female) with extra antennae

24 (10 male + 14 female) with crooked antennae

This is roughly equal to a 50/50 phenotype ratio. It is not perfectly 50/50 because the offspring are generated by random combination of the parent's alleles, as they are in nature. Because of this, small numbers of offspring will only approximate the expected ratios. If you clicked "CROSS" several times for this same pair of parents and added the results together, you would get closer to a 50/50 ratio.

a) You can now select any male and any female insect, find their genotypes, predict their offspring, and test your predictions. Do this for as many crosses as it takes to convince yourself that you understand how this works.

b) As you do this for many crosses, you will see that, although the offspring ratios are not exactly what the Punnett square would predict, it is possible to make some statements about the expected offspring. These statements would use phrases like "none," "all," "roughly equal," and "more than" to describe the expected ratios in these small samples of offspring. For each of the crosses described below, give a phrase that uses one or more of the phrases listed above that accurately describes the expected genotype and phenotype ratios of the offspring.

    i) EE × EE

    ii) EE × Ee

    iii) EE × ee

    iv) Ee × Ee

    v) Ee × ee

    vi) ee × ee

c) Try a new VGL problem.
- From the "File" menu, choose "Close work"; don't save the problem.
- From the "File" menu, choose "New problem."
- From the list of problem files, choose "level1.prb" and click "Open."
- Select "Show model and all genotypes" and click "Continue." You will get a cage with a randomly chosen character. There will be two randomly chosen traits of that character. One is dominant; the other is recessive. Click on the "> show model & genotypes" button to reveal the model.
- Define appropriate symbols for this genetic model.

As you did in part (a), choose a random pair of parents, get their genotypes, predict the expected offspring, cross them, and check your prediction. Keep crossing until you are sure you understand. To try another problem, follow the steps in part (c).

You can save a problem that you are working on by selecting "Save Work As..." from the "File" menu. You can then open this file by choosing "Open Work..." from the "File" menu. All your cages will appear and you can continue working. By default, the program tries to save your files in your home directory; this is not the default directory that opens when you select "Open Work...." Therefore, take careful notice of the folder you save the file in so that you can find it again later.

(1.1.5) What is the simplest explanation that accounts for the following results? Give the genotype of each mouse.

- A brown mouse was crossed with a white mouse, producing 10 brown and 13 white F_1 mice.
- Two white F_1 mice were crossed, giving all white progeny.
- Five pairs of brown F_1 mice were crossed, giving a total of 56 brown and 20 white progeny.

(1.1.6) You are studying coat color in tribbles, a sexually reproducing, diploid species of hypothetical mammals. Tribbles can be either red or blue, and you have other evidence which shows that coat color is determined by a single gene with two alleles (B and b). You cross two tribbles, Tarzan and Jane, and they produce two tribble pups, Fred and Alice:

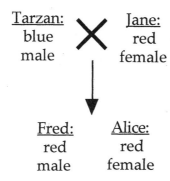

Tarzan: blue male X Jane: red female

Fred: red male   Alice: red female

a) Give a plausible one autosomal gene, two-allele model to explain these data:

  i) Based on your model, give the color phenotype of each of the following genotypes:

Genotype	Phenotype (color)
BB	_____
Bb	_____
bb	_____

ii) Based on your model, give the genotypes of each of the individuals in the following pedigree. Indicate any ambiguities.

Genotype [ ] Tarzan: blue male ✕ Jane: red female Genotype [ ]

↓

Genotype [ ] Fred: red male   Alice: red female Genotype [ ]

b) Give a **different** one autosomal gene, two-allele model for the inheritance of coat color in tribbles that is consistent with the pedigree at the beginning of this problem. To be considered "different," your answer to part b(i) must be different from your answer to part a(i). Parts a(ii) and b(ii) may also be different, but differences in part a(ii) alone do not constitute a "different" model.

    i) Based on your model, give the color phenotype of each of the following genotypes:

Genotype	Phenotype (color)
BB	_____
Bb	_____
bb	_____

ii) Based on your model, give the genotypes of each of the individuals in the pedigree. Indicate any ambiguities.

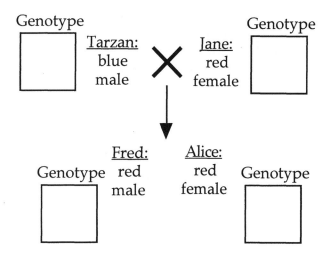

Genotype [ ] Tarzan: blue male ✕ Jane: red female Genotype [ ]

Fred: red male Alice: red female

Genotype [ ] Genotype [ ]

c) You decide to use the rapid reproductive rate of tribbles to distinguish between the two models. You decide to cross two of the tribbles you have and collect many offspring from this cross. You design your cross carefully so that the results of the cross will allow you to determine which is the correct model, (a) or (b).

i) Which two tribbles will you cross? Circle your choices.

Tarzan   Jane   Fred   Alice

ii) What will be the results of the cross of part c(i) if your model from part (a) is true? Give the expected ratio of coat colors.

iii) What will be the results of the cross of part c(i) if your model from part (b) is true? Give the expected ratio of coat colors.

(1.1.7) Recently, some scientists proposed a genetic model to explain the inheritance of left- and right-handedness in humans. Their model is as follows:

Handedness is controlled by one gene with two alleles:

Allele	**Contribution to phenotype**
R	right-handed (dominant)
r	undetermined handedness (recessive)

Therefore:

Genotype	**Phenotype**
RR, Rr	right-handed
rr	undetermined: half of these children develop into left-handed individuals and half develop into right-handed individuals.

a) Based on this model, two Rr parents (right-handed) have a 1:8 chance of having a left-handed child. Explain why this is so.

b) Based on this model, can a left-handed mother and a right-handed father have a left-handed child? Justify your answer.

c) Based on this model, can two left-handed parents have a right-handed child? Justify your answer.

d) One problem with this model is that it is consistent with virtually any combination of left-handed or right-handed parents and offspring. What data, if any, could you imagine finding that would not be consistent with this model?

(V2) Virtual Genetics Lab II Here, you will use VGL in a more advanced mode, and you will have to figure out how the trait is inherited without looking at the answer. This will provide you with more practice with this material. There are many successful ways to approach problems like these; we will begin by looking at a problem that we have already worked through for you.

Launch VGL as before and select "Open Work" from the "File" menu. Select "Problem2.wrk" and click "Open." Be sure that you see a set of several cages. We will go through them in the order that we generated them.

Cage 1 looked like this:

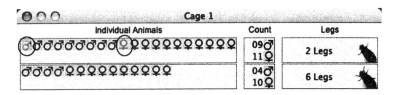

Some flies had two legs and others had six legs. Our task is to determine whether two or six legs is the dominant phenotype.

Given only the two phenotypes (two legs and six legs) and the constraints given for this problem (only one gene; no sex linkage), there are two possible models for the inheritance of this trait. It is good practice to do this at the start of each problem that you work on.

<table>
<tr><td colspan="2"><u>Model A (two legs dominant)</u></td><td colspan="2"><u>Model B: (six legs dominant)</u></td></tr>
<tr><td><u>Phenotype</u></td><td><u>Allele</u></td><td><u>Phenotype</u></td><td><u>Allele</u></td></tr>
<tr><td>two legs (dominant)</td><td>T</td><td>six legs (dominant)</td><td>S</td></tr>
<tr><td>six legs (recessive)</td><td>t</td><td>two legs (recessive)</td><td>s</td></tr>
<tr><td colspan="2">or:</td><td colspan="2">or:</td></tr>
<tr><td><u>Genotype</u></td><td><u>Phenotype</u></td><td><u>Genotype</u></td><td><u>Phenotype</u></td></tr>
<tr><td>TT</td><td>two legs</td><td>SS</td><td>six legs</td></tr>
<tr><td>Tt</td><td>two legs</td><td>Ss</td><td>six legs</td></tr>
<tr><td>tt</td><td>six legs</td><td>ss</td><td>two legs</td></tr>
</table>

The next step is to cross two individuals and look at the resulting offspring to see which model fits. In this problem, we have already done one cross for you: we crossed the two circled insects in Cage 1 and got Cage 2:

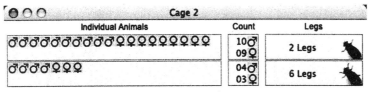

Parent ♂ (1) 2 ♀ (1) 2

There are several important things to notice about Cage 2:
- The bottom line in the Cage window says "Parent Male (1) 2 Female (1) 2." This translates as "the parents are a male from Cage (1) with two legs and a female from Cage (1) with two legs." You can click on the male and female symbols to select those particular parents. This is true for all cages except Cage 1 (which has no parents).
- The results in Cage 2 can be summarized as:

Parents: two legs ✕ two legs
⇓
Offspring: 19 with two legs
             7 with six legs

You will need to use this type of analysis when you are working through other VGL problems as well.

Which model(s) are consistent with the results of the cross that is shown in Cage 2? Starting with Model A, the two-legged parents can be either TT or Tt. TT ✕ TT, Tt ✕ TT, and TT ✕ Tt produce only two-legged offspring, so the parents cannot have these genotypes. However, if the parents were both Tt, they would be expected to produce a 3:1 ratio of two-legged to six-legged offspring. Although 19:7 is not exactly 3:1, given the small sample size, it is close enough for our purposes. In general, when using VGL, the numbers will not come out to exact "Mendelian ratios" because of the small number of offspring. It is therefore useful to think of these ratios in more descriptive terms, as shown below:

Mendelian Ratio (A:B)	Descriptive Phrase
1:0	all A; no B
0:1	all B; no A
1:1	roughly equal amounts of A and B
3:1	more A than B
1:3	more B than A

In the case of Cage 2, we would describe the results as "more two legs than six legs," which is what we would expect from a cross of Tt ✕ Tt.

Next, try Model B. In this case, the two-legged parents must both be ss. Since ss ✕ ss cannot give any six-legged offspring, and we observe six-legged offspring, the data are not consistent with Model B and we can rule it out completely.

Although we were lucky with this cross and this is fairly conclusive, we wanted to be sure we were right. To do this, we used Model A to predict the results of several crosses. We then carried out the crosses to confirm our model.

First, any six-legged insect must be tt, so any pair of six-legged parents will give only six-legged offspring. We crossed two six-legged parents from Cage 2 to produce Cage 3. Since the offspring are all six-legged, this is consistent with Model A.

Our hypothesis is that the original parents were Tt. If that's true, then crossing either of them with a tt (six-legged) insect should give a 1:1 phenotype ratio in the offspring (this corresponds to "roughly equal amounts of two- and six-legged offspring"). The two-legged offspring will all be Tt and the six-legged offspring will be tt. We selected the original male parent by clicking on the male parent symbol at the bottom of Cage 2; we then selected a six-legged female from Cage 3 and crossed them to make Cage 4. The offspring in Cage 4 are:

13 with two legs

14 with six legs

This is a very close match to our prediction, so we are still going strong.

As a last test, we crossed two of the two-legged offspring from Cage 4. Since we expect that these are all Tt, the offspring in Cage 5 should be roughly 3:1 two-legged:six-legged (or, more descriptively, "more two-legged than six-legged"). The observed matches the expected, so we have plenty of evidence to convince ourselves that we have it right.

The most conclusive pieces of evidence are those that are consistent only with one alternative model or the other. It will be useful to keep these in mind when solving VGL problems in the future.

a) Which cages give results that are consistent only with Model A? Why?

b) Which cages give results that are consistent with both Model A and Model B? Why?

c) What other types of results are consistent only with Model A or consistent only with Model B? Why?

d) Make some other predictions based on this model and test them by making crosses.

e) You should now solve a problem on your own. To do this, you:
- Select "Close Work" from the "File" menu. Don't save your work.
- Select "New Problem" from the "File" menu.
- Choose "level2.prb" and click "Open."

You will then have a problem to work through like the one above. One trait will be dominant and the other recessive; your task is to keep crossing until you are convinced which is which. At Level 2, there is no way to see the model or genotypes; you must decide for yourself when you have enough data to be confident in your choice of model.

Hints:
1. Start by writing out the two possible genetic models.
2. Think about what kinds of results would be conclusive (as described above) and keep your eyes open for these.
3. Consider the results of each cross carefully before doing another cross.
4. Once you have a strong suspicion that a particular model is correct, design crosses to test this model as we did above.

You should do several problems this way until you are confident that you understand how to solve them reliably. As you gain more experience, you can use VGL in a manner that more closely approximates the way a geneticist would work in a research lab where the "correct answer" is not known. One possible progression is given below:
- You may want to start by working problems where you can check your work by looking at the model and genotypes as we have shown. For these problems, choose "Level1.prb"; these involve one gene; one trait is dominant to the other.
- As a next step, try to work a Level 1 problem until you are confident that you have found the correct genetic model. Then, and only then, you can check your work by showing the model in Cage 1.
- You can then move on to working problems at Level 2. Here, the genetic models are the same as in Level 1, but you cannot see the model or genotypes. Here, you can work until you are confident in your choice of model and then have a friend check your work. There are several ways to do this:
  - You explain the results of each cage to your friend.
  - Your friend picks any two individuals and you predict the expected offspring. You then do the cross to see if your prediction is correct.

These techniques work for all the different levels of VGL problems. That is, just as the models at Levels 1 and 2 are the same, but Level 2 does not allow you to see the model or genotypes, the possible models at Levels 3 and 4 are the same (but more complex than those at Levels 1 and 2), and Level 4 does not allow you to see the model or genotypes.

(1.2) Pedigrees involving one gene, I

Another way to present genetic data is a pedigree. Pedigrees are useful diagrams for presenting data from many crosses, where each cross generates only a small number of offspring. Pedigrees are a standard tool for evaluating inheritance patterns in humans. Pedigrees use the following symbols to represent the sex and the phenotype of individuals:

■ **Male with the trait of interest. Also called "affected male" in the case of a genetic disease.**

□ **Male lacking the trait of interest. Also called "unaffected."**

● **Female with the trait of interest.**

○ **Female lacking the trait of interest.**

◇ **Unknown sex.**

Relationships between individuals are indicated by lines; as shown, A and B are the parents of C and D.

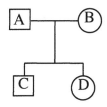

Genetic traits or diseases can be inherited in several different ways. The first two we will consider are:

- <u>Autosomal Dominant</u>: the disease allele gives the *dominant* phenotype.

  Allele Contribution to phenotype

  D   disease (dominant)

  d   normal (recessive)

Genotype	Phenotype	Symbols
DD	affected	■ or ●
Dd	affected	■ or ●
dd	normal	□ or ○

- <u>Autosomal Recessive</u>: the disease allele has the *recessive* phenotype.
    <u>Allele</u> <u>Contribution to phenotype</u>
      N   normal (dominant)
      n   disease (recessive)

Genotype	Phenotype	Symbols
NN	normal	☐ or ○
Nn	normal	☐ or ○
nn	affected	■ or ●

Problems:

(1.2.1)
a) Consider cystic fibrosis, an autosomal recessive genetic disease.
    i) Define appropriate allele symbols for cystic fibrosis.

    ii) Draw a pedigree for a family in which two unaffected parents have a son with cystic fibrosis and an unaffected daughter.

    iii) What is the genotype of the unaffected daughter? Indicate any ambiguity or multiple possibilities.

    iv) What is the chance that the next child in this family will have cystic fibrosis?

b) Consider Marfan syndrome, an autosomal dominant genetic disease.

    i) Define appropriate symbols for Marfan syndrome.

    ii) Draw a pedigree for a family in which both parents have Marfan syndrome and they have an unaffected son and an unaffected daughter.

    iii) What is the genotype of the unaffected son? Indicate any ambiguity or multiple possibilities.

    iv) What is the chance that the next child in this family will have Marfan syndrome?

(1.2.2) For each of the following pedigrees:
- Determine whether the disease is inherited in a dominant or recessive manner.
- Define appropriate allele symbols.
- Give the genotypes of all individuals in the pedigree; be sure to indicate any ambiguities or multiple possibilities. Refer back to page 21 for an explanation of the pedigree symbols.

a)

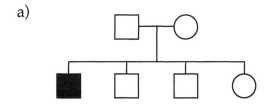

b)

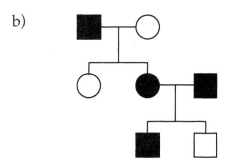

c)

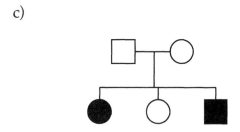

(1.2.3) Although you can solve any pedigree problem by trying all the possible genetic models, this can be rather time-consuming. An alternative method involves looking at pedigrees for particular arrangements of parents and offspring that can rule out one or another mode of inheritance. This can be a more rapid way to eliminate possible modes of inheritance.

For example, one diseased parent and one normal parent having one diseased offspring is consistent with <u>both</u> autosomal dominant and autosomal recessive inheritance. Other arrangements are consistent only with one or the other.

a) For each of the following statements, mark if they are true for autosomal recessive and/or autosomal dominant diseases:

      i) Diseased parents can have diseased offspring.
          • True for autosomal recessive diseases?   Yes   No
          • True for autosomal dominant diseases?   Yes   No

      ii) Normal parents can have normal offspring.
          • True for autosomal recessive diseases?   Yes   No
          • True for autosomal dominant diseases?   Yes   No

iii) Even if both parents are normal, they can have diseased offspring.
- True for autosomal recessive diseases?   Yes   No
- True for autosomal dominant diseases?   Yes   No

iv) Even if both parents are diseased, they can have normal offspring.
- True for autosomal recessive diseases?   Yes   No
- True for autosomal dominant diseases?   Yes   No

b) Can any of these statements be used to rule out one or the other mode of inheritance? Can you apply any of these to problem 1.2.2? How?

(1.2.4) Shown below are two pedigrees for a **rare autosomal recessive** genetic disease. Fewer than 1 in 1,000 people are carriers for this disease.

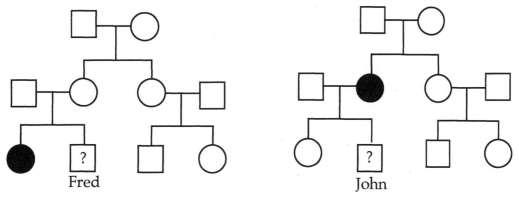

Fred and John are as yet unborn children of parents who are concerned that they may be affected with the genetic disease. Based on the above pedigrees, which individual, Fred or John, has a greater risk of being affected by the disease? Why?

(1.2.5) Read the attached information and consider the pedigree for the family described, then answer the following questions.

    a) Who appears to have Marfan syndrome?

    b) What are the chances that the offspring of the affected individuals will have Marfan syndrome?

Marfan Syndrome Description:

- Inherited as an autosomal dominant trait.
- Defect of connective tissue formation.
- Some common features of the syndrome (not all are present in all cases):
    - Tall stature
    - Long, thin limbs (relative to height)
    - Prominent shoulder blades
    - Spinal curvature (scoliosis)
    - Protruding or caved-in breastbone
    - Flat feet
    - Dislocated eye lens (may require glasses)
    - Detached retina
    - Long fingers and thumbs
    - General joint and cartilage trouble
- Cause of heart valve abnormalities and weakening of the aortal wall. During heavy exercise, the weakened aortal wall can burst, leading to almost instantaneous death.
- No genetic or biochemical test for this syndrome at the time these data were collected.

Information on Family Members

• Anne:

Anne is 16 years old and is a junior in high school. She has read some information in the popular press on Marfan syndrome, and she and her parents are concerned that Anne might have this syndrome. Her general physician has referred her to the Genetics Counseling Clinic. Anne is 5' 11" and wears contact lenses to correct for myopia (nearsightedness). She plays on her school's varsity volleyball and basketball teams. NCAA scouts are already interested in her playing ability, and there is a chance she will be offered college scholarships in both sports. Her armspan/height ratio is 1.08:1. (In one group of 27 adults, this ratio was 1.006 with standard deviation = 0.03.)

Siblings:

• David:

Age 25, married to Jessica, age 25, one daughter named Kristi, age 3 months. David wears glasses, is 6' 3", has long fingers and toes, played basketball and ran track in high school, and had some knee problems that developed during his high school athletic career. Jessica is 5' 8", wears no glasses, and has no health complaints other than occasional migraine headaches. She and David had one miscarried pregnancy in the first trimester before the birth of Kristi.

- Cheryl:
Age 14, 5′ 9″, no glasses, has a slight case of scoliosis. She was born with club feet, which responded well to corrective treatment.

Parents:

- Mary:
Age 47, 5′ 7″, wears glasses, and has hay fever. Has been diagnosed with carpal tunnel syndrome and mild diabetes. Had two miscarriages in addition to her three children.

- Peter:
Age 49, 6′ 1″, wears glasses, concave chest, high blood pressure, partial lens dislocation in left eye, and long fingers and toes. Has complained about chronic tennis elbow.

Aunts and Uncles:

Mary's Siblings:
- Dorothy:
Age 46, wears glasses, 5′ 3″, no major health problems. Had an ovarian fibroid tumor removed at age 40. Married and has four children.

- Ellen:
Age 50, 5′ 5″, high cholesterol, has been diagnosed with irritable bowel syndrome. Unmarried, no children.

- Eric:
Age 51, 6′ 1″, wears reading glasses, has recurrent back problems from a car accident, suffers from exercise-induced asthma. Is married and has two children from his first marriage and three from his second.

Peter's Siblings:
- Frank:
Age 55, 6′ 4″, wears glasses, slight hearing loss in one ear. Was treated for alcoholism, is a heavy smoker, and has developed a chronic cough. Divorced, the father of two children.

- Alice:
Age 56, 5′ 7″, wears glasses, arthritis in left shoulder. Married, has one daughter and a son who was born with cerebral palsy.

- John:
Deceased, heart attack at age 46, 6′ 2″, had dislocated lens in right eye. He and his wife had three children. Their youngest daughter is slightly mentally retarded.

- Larry:
Age 58, 6′ 3″, no glasses, high blood pressure. Divorced twice, lives alone now. Had two children by his first marriage and one by his second. Is a heavy drinker.

Anne's Maternal Grandparents:
• Evelyn:
Died at age 76 of stroke, 5' 4", arthritis in hands and feet, wore reading glasses. Was said to have had as many as five miscarriages.

• William:
Age 81, no glasses, 5' 10", no major health problems. Has a slight limp due to bad right knee, occasional rashes, and hemorrhoids.

Anne's Paternal Grandparents:
• Gloria:
Age 86, 5' 8", high blood pressure, some knee and ankle problems. Is concerned about her constipation. Wears glasses for distance and reading.

• Charlie:
Died at age 44 of a heart attack, severe vision problems, described as long and lanky. Contracted polio at age 26. He was wheelchair dependent following polio treatment.

Anne's Family Tree:

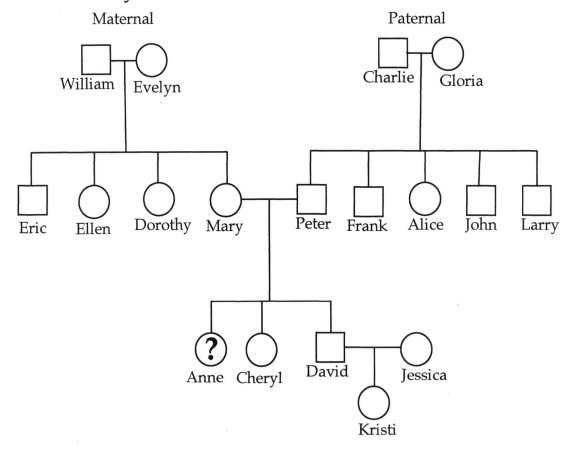

Chapter 1: Genetics Problems

(1.3) One gene; more complex models, I

In this section, we will deal with more complex single-gene models. The models are more complex because i) they do not follow simple dominance or ii) there are more than two alleles for the gene.

a) Two alternatives to simple dominance are incomplete dominance and codominance. To evaluate these more complex models, let's first review simple dominance. So far, we have evaluated a single gene trait with two alleles in the following way. We could have homozygous organisms with the genotypes:

Genotype	Phenotype
AA	red
aa	blue

If the heterozygote (genotype Aa) is red, then red is the dominant phenotype and blue is the recessive phenotype.

In *incomplete dominance*, the heterozygote has an *intermediate* phenotype, a phenotype different from either of the homozygotes. For example, we could have homozygous organisms with the genotypes:

Genotype	Phenotype
TT	red
T'T'	blue
TT'	purple, this is intermediate between red and blue

In *codominance*, the heterozygote has a *mixture of both* homozygote phenotypes. For example:

Genotype	Phenotype
TT	red
T'T'	blue
TT'	a mixture of both red and blue

In both cases, the heterozygote can be distinguished from either of the homozygotes. Lowercase letters were not used as allele symbols because the lowercase letters symbolize alleles associated with a recessive phenotype.

b) The complexity of genetic models increases when you have more than two alleles of a gene because the number of possible genotypes increases. If a gene has three alleles, then six possible genotypes exist. A classic example of a gene in which there are three alleles is the blood-type gene in humans. The three blood types are A, B, and O, so it might make sense to use A, B, and O as the allele symbols. This creates a problem because the genotypes and the phenotypes are then easily confused. For that reason, we will use symbols like these when dealing with more than two alleles of the same gene: I^A, I^B, or i.

Blood type in humans is a good example problem for this because it shows both simple dominance and codominance.

Example problem

With respect to blood type in humans, all possible genotypes are given below.

Genotype	Phenotype
$I^A I^A$	type A
$I^B I^B$	type B
ii	type O
$I^A i$	type A
$I^B i$	type B
$I^A I^B$	type AB

a) Given this information, circle all the true statements.

Type A blood is dominant to type B blood.

Type A blood is dominant to type O blood.

Type O blood is recessive to blood types A and B.

Type A and type B blood types are codominant.

Type B blood is codominant to type A blood and dominant to type O blood.

Type B blood is incompletely dominant to type A blood but dominant to type O blood.

b) A dad with type B blood and a mom with type A blood have a child with type O blood. Give the genotypes of the two parents.

Answers to example problem

a) Given this information, circle all the true statements.

Type A blood is dominant to type B blood.

Type A blood is dominant to type O blood.

Type O blood is recessive to blood types A and B.

Type A and type B blood types are codominant.

Type B blood is codominant to type A blood and dominant to type O blood.

Type B blood is incompletely dominant to type A blood but dominant to type O blood.

b) A dad with type B blood and a mom with type A blood have a child with type O blood. Give the genotypes of the two parents.

Dad must be $I^B i$.
Mom must be $I^A i$.

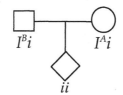

Problems:

(1.3.1) Suppose that the height of a hypothetical plant is controlled by one gene with two incompletely dominant alleles.

a) What phenotypes would be expected in the offspring from a cross of a tall plant and a short plant?

b) What phenotypes would be expected in the offspring from a cross of two medium plants?

(1.3.2) Suppose that the hair length of a hypothetical mammal is controlled by one gene with two codominant alleles.

a) What phenotypes would be expected in the offspring of the cross of a long-haired critter and a short-haired critter?

b) What phenotypes would be expected in the offspring of a cross of two critters with mixed long and short hair?

(1.3.3) Suppose that color in a hypothetical flower is controlled by one gene with three alleles. Assume that blue color is dominant to all, and red color is dominant to green, but recessive to blue. Using the alleles C^B, C^R, and c, construct a table of all six possible genotypes and the corresponding phenotypes.

(1.3.4) Consider the flower color in a hypothetical plant. Make a genetic model that fits the following data and give the genotypes of the different groups of individuals.

Cross 1: Blue-flowered plant ✕ white-flowered plant

gives F_1:   all pale-blue-flowered

Cross 2: Pale-blue F_1 ✕ pale-blue F_1

gives F_2:   27 blue
49 pale-blue
24 white

(1.3.5) You are studying eye color in an imaginary fly. You know that red, white, and green eye colors are seen and that eye color is controlled by a single gene.

Cross 1:
You cross two green-eyed flies and get some green-eyed and some white-eyed offspring.

a) Given only cross 1, generate the simplest genetic model that fits the data.
    i) Define your allele symbols clearly.

    ii) What are the genotypes of the two parents of cross 1?

    iii) What is/are the genotype(s) of the green-eyed offspring?

Cross 2:
    red-eyed ✗ white-eyed gives half red-eyed and half green-eyed offspring

b) Now give a genetic model that fits **all the data from both crosses**:
Note: your answers to parts (a) and (b) need not be the same.

    i) Define your allele symbols clearly.

    ii) What are the genotypes of the two parents of **cross 2**?

(1.3.6) You're the director of Beth Israel Hospital's maternity ward, and you have four mixed-up babies, with blood types of O, A, B, and AB. The four sets of desperate parents are threatening to sue you if you don't get their babies back. You know the blood types (phenotypes) of the parents are:

   i) AB and O
   ii) A and O
   iii) A and AB
   iv) O and O

You also know that each set of parents has only one child (no sets of twins, that is). Match each baby with its parents.

(1.3.7) You solved the baby mystery with flying colors and received a commendation from the hospital director for brilliance. Two days later, however, you are faced with an even more complicated situation: a male (**George, type B blood**) and a female (**Sallie, type A blood**) claim that a newborn (**Fred, type B blood**) is their son.

a) Given this information, is it possible that George and Sallie are Fred's parents? (explain briefly)

b) You learn that George's father has type A blood and his mother has type B blood. Given this information, is it possible that George and Sallie are Fred's parents? (explain briefly)

c) On further investigation, you find that George's sister has type O blood. Given this information, is it possible that George and Sallie are Fred's parents? (explain briefly)

d) Finally, you discover that both of Sallie's parents are type AB. Given this information, is it possible that George and Sallie are Fred's parents? (explain briefly)

(1.3.8) A woman with blood type O has a child with blood type O. She claims that her friend, Bob, is the child's father.

a) Bob's blood type is A. Can he be excluded as the father on this evidence alone?

b) Does the fact that Bob's mother has type A blood and his father has type AB blood exclude him from being the parent?

c) Does the additional information that Bob's mother's parents are both AB permit him to be excluded?

(1.3.9) You are working in the maternity ward of a major Boston hospital. There are three babies in your care:

Baby	Blood type
Cathy	A
Steve	B
Rodger	O

One of the doctors has lost all the information matching the babies with their parents. Your job is to match the babies with their parents. So far, you have blood-type data on only two couples:

- Couple #1: **Tom** (type AB) and **Ann** (type A)
- Couple #2: **Peter** (type B) and **Sally** (type B)

The other couple has not yet given you blood samples. With the data you have, you try to match the babies and parents.

a) Given the information on the previous page, can you <u>rule out</u> any of the couples as the parents of any of the babies? Explain your reasoning.

b) You learn that Ann's parents have blood types AB and AB. Based on the information you have <u>so far</u>, can you exclude any of the couples as the parents of any of the babies? Explain your reasoning.

c) You learn that Peter's parents have blood types B and O and that Sally's parents have blood types AB and AB. Based on the information you have <u>so far</u>, can you exclude any of the couples as the parents of any of the babies? Explain your reasoning.

(V3) Virtual Genetics Lab III In this section, you will work VGL problems that include a larger set of possible models. All involve one gene with two or three alleles and a maximum of three different phenotypes. The problems involve the following four models:

- *Two alleles with simple dominance* – this will yield two different phenotypes. You have worked with problems of this type before; we have included them in this set to make you stretch your knowledge of genetics a little.
- *Two alleles with incomplete dominance* – this will yield three different phenotypes: homozygote 1, homozygote 2, and the heterozygote. Because VGL chooses traits randomly, the heterozygote's phenotype may not be intermediate between the two homozygotes (for example, six legs may be the heterozygote of four legs and two legs).
- *Three alleles with "hierarchical" dominance* – this will yield three different phenotypes. In this case, A^1 is dominant to A^2 and A^3, A^2 is dominant to A^3 and recessive to A^1, and A^3 is recessive to all. It is not clear that situations such as these occur in nature, but we have included them to help you explore more advanced genetic models.
- *Three alleles with "circular" dominance* – this will yield three different phenotypes. In this case, B^A is dominant to B^B, B^B is dominant to B^C, and B^C is dominant to B^A. It is not clear that situations such as these occur in nature, but we have included them to help you explore more advanced genetic models.

We have generated a VGL problem with each of these types of inheritance so that you can explore the differences between these models. To run them, double-click on the VGL icon in the "Genetics" folder. Then choose "Open Work..." from the File menu and select one of the following files:

- *Two alleles with incomplete dominance* Problem3_ID.wrk In this case:
  - $B^B B^B$ = blue body color
  - $B^P B^P$ = purple body color
  - $B^B B^P$ = pink body color
- *Three alleles with "hierarchical" dominance* Problem3_HD.wrk In this case:
  - $B^G B^G$ = green body color
  - $B^R B^R$ = red body color
  - $B^P B^P$ = purple body color
  - $B^G B^R$ = green body color
  - $B^G B^P$ = green body color
  - $B^R B^P$ = red body color
- *Three alleles with "circular" dominance* Problem3_CD.wrk In this case:
  - $E^G E^G$ = green eye color
  - $E^B E^B$ = blue eye color
  - $E^Y E^Y$ = yellow eye color
  - $E^G E^B$ = blue eye color
  - $E^B E^Y$ = yellow eye color
  - $E^G E^Y$ = green eye color

Be sure to click "Show model and all genotypes" and click "Continue."

Spend some time working with each model to be sure that you understand the differences between them.

Now that you are familiar with these models, you should try a few problems where you have to determine the model yourself.

As a reminder, since VGL selects characters and traits randomly, the particular traits do not necessarily indicate the dominance relationships. That is, although you might expect otherwise, having no antennae may be dominant to having antennae. Similarly, having four legs may not be the heterozygote of two legs and six legs.

You should solve several VGL problems at this level; keep at it until you are sure that you understand the differences between simple, incomplete, hierarchical, and circular dominance and how to tell them apart.

- Double-click the VGL icon in the "Genetics" folder.
- Choose "New Problem" from the "File" menu.
- Select "level3.prb" to start; it allows practice mode. Try not to look at the "answer" at first.
- Solve the problem.
- Have a friend check your work. Your friend picks any two insects and you predict the expected offspring. You then do the cross to see if your prediction is correct. You can also move up to Level 4, which has the same genetic models but does not allow you to look at the answer.

(1.3.10) You are studying a strange (and hypothetical) population of alien plants; these plants are diploid. You find plants in two colors: purple and blue. You do some crosses to find out how color is inherited.

Cross 1		**Cross 2**
P: purple ✕ blue		**P:** purple ✕ blue
⇓		⇓
F₁: 247 purple		**F₁:** 460 purple
262 blue		

Where F₁ is F_1.

a) <u>Based on this information only</u>, give a genetic model of color inheritance in these plants. Define appropriate allele symbols and give their contribution to phenotype:

You find some red plants and do some more crosses:

Cross 3	**Cross 4**	**Cross 5**
purple ✕ purple	purple ✕ purple	blue ✕ red
⇓	⇓	⇓
3:1 purple:blue	3:1 purple:red	all blue

b) Based on all the information provided in this question, give a genetic model of color inheritance in these plants. Define appropriate allele symbols and give their contribution to phenotype:

c) Using the symbols you defined in part (b), give the genotypes of the following parents. If more than one genotype is possible, <u>give all possibilities</u>.

- Cross 1   purple parent   _____

- Cross 2   purple parent   _____

- Cross 3   purple parent   _____

- Cross 4   purple parent   _____

- Cross 5   blue parent   _____

(1.3.11) A zoologist friend of yours has just discovered a new creature whose most striking characteristic is its vivid coat colors. She has promptly named the creature *Marge tribblicus,* commonly known as the tribble. Tribbles come in coat colors of pink, green, red, or white. Your friend is interested in understanding the inheritance of tribble coat colors and asks you for your help. She gives you pure-breeding green, red, and white tribbles. You set up the following crosses with them:

Cross 1:   P: one green tribble **✕** one red tribble
produced: F_1:   10 green tribbles

Cross 2:   P: one green tribble **✕** one white tribble
produced: F_1:   10 green tribbles

After these results you plan cross 3 where you cross the green F_1 tribbles from cross 1 with the green F_1 tribbles from cross 2:

Cross 3:   P: green F_1 tribble from cross 1 **✕**   green F_1 tribble from cross 2
produced: F_1:   31 green tribbles and 10 pink tribbles

a) What is the simplest explanation for the results from these three crosses? Give the genotypes of each class of progeny.

b) Predict the results of a cross of the pink F_1 tribble from cross 3 and a pure-breeding green tribble.

(1.4) One gene; sex linkage

In this section you will explore problems that follow simple dominance and have two alleles for the gene of interest. What is unique about these problems is that the gene of interest is carried on one of the sex chromosomes. All diploid organisms have two copies of each chromosome, the exception being the sex chromosomes. In mammals, females have two copies of the X chromosome but males have only one copy of the X chromosome and one copy of the Y chromosome. Different organisms have alternative sex determination systems. In birds, males have two of the same sex chromosomes and females have two different sex chromosomes. Male birds have the genotype, with respect to the sex chromosomes, of ZZ. Female birds have the genotype, with respect to the sex chromosomes, of ZW.

The inheritance pattern of a single gene trait carried on a sex chromosome differs from the autosomal inheritance patterns that we have seen so far. For example, in the following fly cross:

Female with short wings	✗	Male with long wings
ll		LL

↓

| 50% | Ll | Females with long wings |
| 50% | Ll | Males with long wings |

Because all offspring get one chromosome from each parent, they all get an L allele from dad and they all show the dominant phenotype.

But in a cross where the gene of interest is on the X chromosome, we see a different outcome. Note the standard allele symbols designating X chromosome linkage.

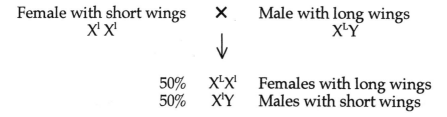

Female with short wings	✗	Male with long wings
$X^l X^l$		$X^L Y$

↓

| 50% | $X^L X^l$ | Females with long wings |
| 50% | $X^l Y$ | Males with short wings |

All offspring get one sex chromosome, X^l, from mom. They also must get a sex chromosome from dad, but it can be either X^L or Y. A fly that gets the X^L from dad is female and shows the dominant phenotype. A fly that gets Y from dad is male and shows the recessive phenotype because he has only one allele for the wing length trait, the X^l allele.

With traits carried on the X chromosome (X-linked traits), males will tend to show recessive traits more frequently than females. In general, you should suspect sex linkage if you see a result where particular phenotypes are not equally distributed between the different sexes. For example, this result:

#	Phenotype
25	red male
22	green male
26	red female
24	green female

does not suggest sex linkage because there are roughly equal numbers of red and green males and females. This does not mean that sex linkage is not involved, just that this result, on its own, does not provide evidence for sex linkage. On the other hand, this result:

#	Phenotype
25	red male
27	green female

strongly suggests sex linkage. Although there are both red and green offspring and both males and females, you see that there are no green males and no red females – the phenotypes are not distributed equally by sex.

Problems

(1.4.1) Consider the following X-linked trait in a hypothetical mammal with XX/XY sex-determination where red eyes are dominant to white eyes.

a) Predict the expected offspring from the following crosses:

i) White-eyed female ✕ red-eyed male.

ii) Red-eyed female ✕ white-eyed male (there are two possibilities here; give both).

Consider the following Z-linked trait in a hypothetical bird with ZZ/ZW sex determination where red eyes are dominant to white eyes.

b) Predict the expected offspring from the following crosses:

i) White-eyed female **✗** red-eyed male (there are two possibilities here; give both).

ii) Red-eyed female **✗** white-eyed male.

(1.4.2) There are now four possible modes of inheritance to consider when solving human pedigrees:
- Autosomal recessive (AR)

- Sex-linked recessive (SLR)

- Autosomal dominant (AD)

- Sex-linked dominant (SLD)

In most of the problems in this book, we will not consider sex-linked dominance. This is because sex-linked dominant diseases are very rare, but, more important, it is very difficult to tell a sex-linked dominant pedigree from an autosomal dominant one (see later).

As we have said before, when working pedigrees, it is often useful to have certain diagnostic configurations that can be used to rule out one or more modes of inheritance. For each of the three types of inheritance listed below, give a combination of parents (either diseased or normal) that give offspring (either diseased or normal) that are <u>inconsistent</u> with that mode of inheritance.

a) A combination that is inconsistent with autosomal recessive.

b) A combination that is inconsistent with sex-linked recessive.

c) A combination that is inconsistent with autosomal dominant.

(V4) Virtual Genetics Lab IV In these two problems, you will work with XX/XY and then ZZ/ZW sex linkage.

a) In this VGL problem, eye color is controlled by a gene located on the X chromosome. In these insects, females are XX and males are XY, and green eyes are dominant to yellow eyes.

- Launch VGL and select "Open Work" from the "File" menu.
- Choose "Problem4_XX.XY.wrk."
- Select "Show model and all genotypes" and click "Continue."
- Click on the ">Show model & genotypes" button.

Select two insects, get their genotypes, predict the expected offspring, and check your prediction. Keep doing this until you are sure you understand the differences between this and autosomal inherited characters.

b) In this VGL problem, body color is controlled by a gene located on the Z chromosome. In these insects, females are ZW, males are ZZ, and pink body is dominant to yellow body.

- Launch VGL and select "Open Work" from the "File" menu.
- Choose "Problem4_ZZ.ZW.wrk."
- Select "Show model and all genotypes" and click "Continue."
- Click on the ">Show model & genotypes" button.

Select two insects, get their genotypes, predict the expected offspring, and check your prediction. Keep doing this until you are sure you understand the differences between this and autosomal inherited characters.

(1.5) Pedigrees involving one gene, II

In analyzing pedigrees, the objective is to produce a genetic model that fits the data given. Assume the trait is associated with one gene with two alleles, one disease allele and one normal allele. A complete model includes the following two components.

1) The mode of inheritance: autosomal recessive, sex-linked recessive, or autosomal dominant.

* Note that there are two ways that questions regarding the mode of inheritance can be phrased:

Which modes of inheritance are consistent with this pedigree? Of the three modes of inheritance, which can explain this pedigree? There can be more than one answer to this question.

What is the most likely mode of inheritance? If more than one mode is consistent with the pedigree, then it is sometimes possible to decide which of the consistent modes is more likely. To evaluate which mode of inheritance is more likely, determine the number of unrelated carriers that are required for each mode. If you can assume that unrelated carriers are rare, the mode that requires the fewest unrelated carriers is the more likely.

2) The genotypes of all individuals in the pedigree.

If the trait is rare, individuals who marry into the family are unlikely to be carriers. We also assume that rare events like nondisjunction and mutation do not occur.

(1.5.1) For each of the following pedigrees:
- What is the most likely mode of inheritance for this trait?
- Give the genotypes of all individuals in this pedigree.
- Suppose the couple indicated with an asterisk have another son. What is the chance that he will be affected?

Key to Symbols:

□ = normal male   ○ = normal female

■ = affected male   ● = affected female

a)

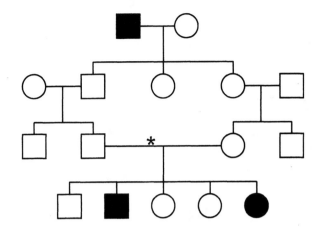

b)

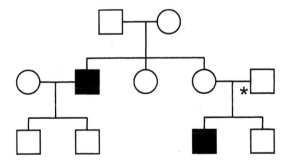

(1.5.2) Consider the following pedigree for a **rare** genetic trait:

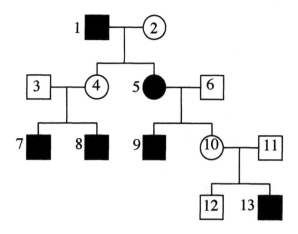

a) What is the most likely mode of inheritance for this trait?

b) Based on your answer to part (a), define appropriate symbols and give the genotypes of all the members of this family.

(1.5.3) Consider the following pedigree for a rare human genetic trait.

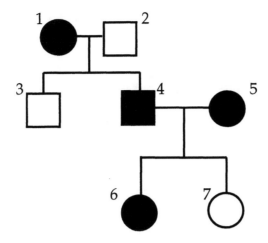

a) What is the most likely mode of inheritance of this trait?

b) Based on your model of part (a), define appropriate allele symbols.

c) Using your symbols from part (b), give the genotypes of all members of the family.

d) If 4 and 5 have another daughter, what is the chance that this daughter will be affected? Justify your answer.

(1.5.4) Hemophilia is a genetic disease that causes affected individuals to have difficulty forming blood clots. It is inherited as an X-linked recessive trait. Apparently, certain ancient cultures were aware of this. For example, the Talmud instructs Jews to perform circumcision (surgical removal of the foreskin) on all males **except** those whose mother's brother is a "bleeder" (hemophiliac).

a) Does this exemption make sense in genetic terms? Why or why not? Explain, using a pedigree.

b) There is no similar exemption for a son whose father's brother is a bleeder. Is this an oversight, or does it make sense in genetic terms? Why or why not? Explain, using a pedigree.

c) Should an exemption be made for the son of a mother whose father is a bleeder? Explain.

(1.5.5) For the following human pedigrees, determine:
    i) The most likely mode of inheritance.
    ii) The probable genotype of the individual marked with an asterisk (*).
Assume that the disease allele is rare. "Rare" means that individuals who marry into the family are very unlikely to have the defective allele. Explain your reasoning and include any ambiguities. Be sure to define your genotype symbols clearly.

a)

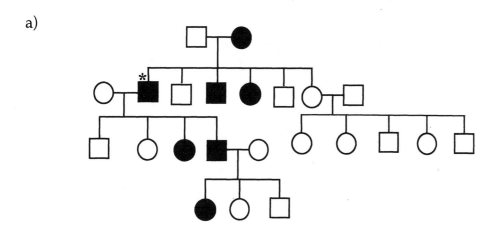

b)

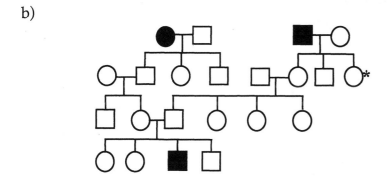

c)

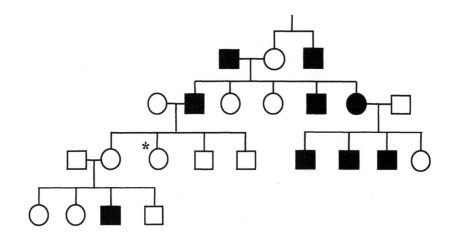

d)

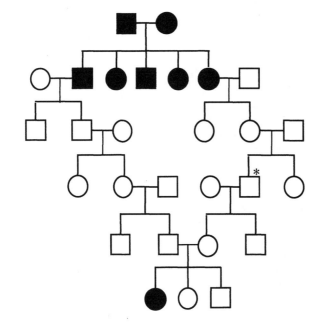

Chapter 1: Genetics Problems

e)

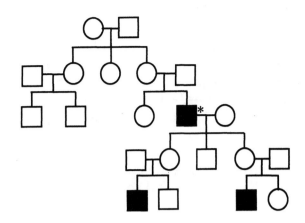

f)

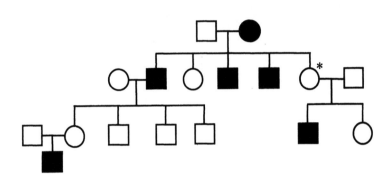

g)

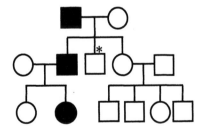

(1.6) One gene; more complex models, II

(V5) Virtual Genetics Lab V In this section, you will work VGL problems that include an even larger set of possible models. All involve one gene with two alleles and only simple dominance. They may involve autosomal or sex linkage. If they are sex-linked traits, they can be either XX/XY or ZZ/ZW. Your task is to find out:

1) Is this trait XX/XY sex linked, ZZ/ZW sex linked, or autosomal?
2) Which trait is dominant and which is recessive?

You should solve several VGL problems at this level; keep at it until you are sure that you understand the differences between autosomal and sex-linked traits and how to tell them apart.

- Double-click the VGL icon in the "Genetics" folder.
- Choose "New Problem" from the "File" menu.
- Select "level5.prb" to start; it allows practice mode. Try not to look at the "answer" at first.
- Solve the problem.
- Have a friend check your work. Your friend picks any two insects and you predict the expected offspring. You then do the cross to see if your prediction is correct. You can also move up to Level 6, which has the same genetic models but does not allow you to look at the answer.

(V6) Virtual Genetics Lab VI These are the most challenging VGL problems. There are 12 possible genetic models. All the problems in VGL involve genetic models with one gene that has two or three alleles. Based on this, there are several features that can vary:

- The number of alleles; this can be either:
  - Two alleles (Models 1, 2, 3, 4, 5, and 6). Given this, there are two possible **interactions between the alleles**:
    - Simple Dominance (Models 1, 3, and 5). The heterozygote has the same phenotype as the dominant homozygote.
    - Incomplete Dominance (Models 2, 4, and 6). The heterozygote has a different phenotype than either homozygote. In nature, this is usually intermediate; in VGL it need not be.
  - Three alleles (Models 7, 8, 9, 10, 11, and 12). Given this, there are two possible **interactions between alleles**:
    - Hierarchical Dominance (Models 7, 9, and 11). A is dominant to all; A' is dominant to A'' and recessive to A; A'' is recessive to all (A > A' > A'').
    - Circular Dominance (Models 8, 10, and 12). B is dominant to B'; B' is dominant to B''; B'' is dominant to B.
- Whether the trait is sex linked or not; this can be either:
  - Not sex linked (Models 1, 2, 7, and 8). The gene for the character is carried on an autosome so it is inherited identically in both sexes.
  - Sex linked. The gene for the trait is located on a sex chromosome so it is inherited differently in different sexes. **This can be either**:
    - XX/XY (Models 3, 4, 9, and 10). Females are XX; males are XY. Here, Y carries no genes except those needed to make the organism male.
    - ZZ/ZW (Models 5, 6, 11, and 12). Females are ZW; males are ZZ. Here, W carries no genes except those needed to make the organism female.

This leads to six possible genetic models.
- Model 1: 2 alleles; Simple Dominance; Autosomal.
- Model 2: 2 alleles; Incomplete Dominance; Autosomal.
- Model 3: 2 alleles; Simple Dominance; XX/XY Sex linked.
- Model 4: 2 alleles; Incomplete Dominance; XX/XY Sex linked.
- Model 5: 2 alleles; Simple Dominance; ZZ/ZW Sex linked.
- Model 6: 2 alleles; Incomplete Dominance; ZZ/ZW Sex linked.
- Model 7: 3 alleles; Hierarchical Dominance; Autosomal.
- Model 8: 3 alleles; Circular Dominance; Autosomal.
- Model 9: 3 alleles; Hierarchical Dominance; XX/XY Sex linked.
- Model 10: 3 alleles; Circular Dominance; XX/XY Sex linked.
- Model 11: 3 alleles; Hierarchical Dominance; ZZ/ZW Sex linked.
- Model 12: 3 alleles; Circular Dominance; ZZ/ZW Sex linked.

Your task is to find the genetic model that best fits your data.

You should know that, since VGL selects traits randomly, the particular traits do not necessarily indicate the dominance relationships. That is, although you might expect otherwise, having no antennae may be dominant to having antennae. Similarly, having four legs may not be the heterozygote of two legs and six legs.

You should solve several VGL problems at this level; keep at it until you are sure that you understand the differences between the models and how to tell them apart.
- Double-click the VGL icon in the "Genetics" folder.
- Choose "New Problem" from the "File" menu.
- Select "level7.prb" to start; it allows practice mode. Try not to look at the "answer" at first.
- Solve the problem.
- Have a friend check your work. Your friend picks any two insects and you predict the expected offspring. You then do the cross to see whether your prediction is correct. You can also move up to Level 8, which has the same genetic models but does not allow you to look at the answer.

(2) PROBLEMS INVOLVING TWO OR MORE GENES

These problems require you to extend what you have learned about one gene to situations with more than one gene. In these problems, it is especially useful to specify different letters (symbols) for each gene whenever possible.

(2.1) Two or more genes that assort independently (Mendel's Second Law)

Work through the diagnostic question on your own and then look at our approach to solving it. If any of the underlined terms are unfamiliar, please consult your book's chapter on di-hybrid crosses.

Diagnostic Question:
You are following two single-gene traits in peas: seed color and seed shape. Green seeds are dominant to yellow seeds, and smooth seeds are dominant to wrinkled seeds. You cross a true-breeding pea plant that has smooth green seeds with a pea plant that has wrinkled yellow seeds.

a) Give the genotypes for each parent.

b) Show the Punnett square for this cross.

c) In a di-hybrid cross of two F_1 plants, what phenotypic ratio do you expect in the offspring?

Answer to Diagnostic Question:

Some appropriate symbols would be:

Allele	contribution to phenotype
GG or Gg	green seeds (dominant)
gg	yellow seeds (recessive)

Genotype	phenotype
SS, Sr	smooth seeds (dominant)
ss	wrinkled seeds (recessive)

Remember that we are following two traits in a <u>diploid</u> organism. Thus, we will represent each parent with four alleles.

a)   smooth green seeds   ✗   *wrinkled yellow seeds*
                SSGG   ssgg

b) *The rows and columns in the Punnett square correspond to <u>gametes</u> made by each parent. Each gamete has one copy of each gene. Therefore, SG is a possible gamete (it has one copy of the seed color gene and one copy of the seed shape gene) but GG is not (it has two copies of the seed color gene and no copies of the seed shape gene). The <u>true-breeding</u> plant with smooth green seeds has two copies of the S allele and two copies of the G allele ($S_1S_2G_1G_2$); the possible gametes are S_1G_1, S_1G_2, S_2G_1, and S_2G_2. Thus, the Punnett square is:*

	S_1G_1 or SG	S_1G_2 or SG	S_2G_1 or SG	S_2G_2 or SG
sg	SsGg	SsGg	SsGg	SsGg
sg	SsGg	SsGg	SsGg	SsGg
sg	SsGg	SsGg	SsGg	SsGg
sg	SsGg	SsGg	SsGg	SsGg

All plants from this cross have the genotype SsGg and have smooth green seeds.

c) *The F_1 plants have the genotype SsGg. Each plant will produce four different types of gametes: SG, Sg, sG, and sg. Thus, the Punnett square is:*

	SG	Sg	sG	sg
SG	SSGG Smooth green	SSGg Smooth green	SsGG Smooth green	SsGg Smooth green
Sg	SSGg Smooth green	SSgg Smooth yellow	SsGg Smooth green	Ssgg Smooth yellow
sG	SsGG Smooth green	SsGg Smooth green	ssGG Wrinkled green	ssGg Wrinkled green
sg	SsGg Smooth green	Ssgg Smooth yellow	ssGg Wrinkled green	ssgg Wrinkled yellow

The ratio is 9 smooth green to 3 smooth yellow to 3 wrinkled green to 1 wrinkled yellow.

Problems

(2.1.1) Consider a hypothetical flowering plant that can have red or green flowers and tall or short stems. For flower color, green is dominant to red, and tall stems are dominant to short stems.

Predict the phenotypic ratios of offspring from the following crosses:

a) Green tall (GGTT) ✕ red short (ggtt).

b) Green tall (GgTt) ✕ red short (ggtt).

c) Green short (Ggtt) ✕ red tall (ggTt).

d) Green short (GGtt) ✕ green tall (GgTt).

e) Green tall (GgTt) ✕ green tall (GgTt).

(2.1.2) You are trying to breed a particularly unusual hypothetical fly. The three traits you are working with are:

eye color: red or white
wing shape: straight or curly
body shape: normal or tubby

If you cross

red-eyed		red-eyed
curly-winged	✗	straight-winged
normal body		tubby body

Your F_1 generation is:
75%   curly tubby red eyes
25%   curly tubby white eyes

a) Assuming that these three traits are each determined by a single gene and the three genes are autosomal and not linked, which is the dominant and which is the recessive phenotype for each trait?

b) What are the genotypes of the two parents in the cross above? Be sure to define your genotype symbols clearly.

c) If two white-eyed, curly-winged, tubby F_1's cross, what will be the ratio of progeny in the F_2?

(2.1.3) Intrigued by the genetics covered in this book and trapped indoors by a severe blizzard, you decide to study the genetics of the flies in your kitchen. You find that the flies have two possible eye colors: red or black; and two possible body colors: green or brown. NOTE: Because these are not lab flies, you should not assume that they are true breeding.

You set up some crosses to study the inheritance of these traits.
Cross 1:

a red-eyed, brown-body fly ✕ a red-eyed, brown-body fly

gave: 42 red-eyed , brown-body progeny

This didn't seem very interesting, so you tried crossing flies with another phenotype:
Cross 2:

a black-eyed, green-body fly ✕ a black-eyed, green-body fly

gave: 32 black-eyed, green-body flies
11 red-eyed, green-body flies

The varied progeny encouraged you to try another cross, using flies different from the ones you used in cross 2:
Cross 3:

a black-eyed, green-body fly ✕ a black-eyed, green-body fly

gave: 10 black-eyed, brown-body flies
31 black-eyed, green-body flies

Finally, you try:
Cross 4:

a black-eyed, green-body fly ✕ a red-eyed, brown-body fly

gave: 32 black-eyed, green-body flies

a) Which phenotypes are dominant and which are recessive?

b) Define the genotype symbols for these alleles (remember to use one letter per gene).

c) Show how your model of part (a) fits the data from the four crosses by giving the genotypes of the parents and offspring in each cross.

(2.1.4) You decide to study the genetics of a hypothetical houseplant. You are studying two traits. You may assume that each trait is the result of a single gene.

      Leaf shape: broad or narrow
      Stem height: long or short

You are given several dozen seeds from the mating of two different true-breeding parent plants, but you do not know the phenotypes of the parents. You grow the seeds into adult plants and find that they all have the same phenotype: broad leaves, long stems. Call these plants the F_1 generation. Now, you allow the F_1 plants to cross-pollinate. You collect seeds, grow them into adult plants, and count phenotypes. Call these plants the F_2 generation.

      Your results:
          F_2 phenotypes

Broad leaves, long stem	28
Broad leaves, short stem	10
Die as seedlings	9
Narrow leaves, short stem	3

a) Which leaf phenotype is dominant? Which stem phenotype is dominant? Explain.

b) What must be the genotype of the F_1 plants? Be sure to define your allele designations.

c) What phenotype class is missing in the F_2 generation? Provide a plausible explanation as to why this class is missing.

d) What must have been the phenotype and genotype of the two parents of your original seeds?

(2.2) Two or more genes that are linked

These problems involve a violation of Mendel's law of independent assortment. In this case, two (or more) genes are positioned close together on the same chromosome so that they do not assort independently. During meiosis, chromosome pairs divide and one of each pair ends up in a different gamete. Genes that are physically near to one another on the same chromosome are likely to end up in the same gamete. Genes that do not assort independently because they are positioned close together on the same chromosome are called <u>linked</u> genes.

When working problems with linked genes, we use a modified set of symbols that shows both an individual's alleles and the specific chromosomes that carry those alleles. There are many possible symbols, but we will use those based on the allele symbols we have used in the past.

Suppose that we have two genes, A and B, each with two alleles. Suppose further that these genes are on the same chromosome. Consider an individual with the genotype AaBb. If the genes are unlinked (assort independently), this individual will produce four types of gametes (AB, Ab, aB, and ab) at equal frequencies. If the genes are close together on the same chromosome (i.e., the genes are linked), then frequencies of the four different types of gametes this individual produces will depend on the recombination frequency <u>and</u> which alleles are on the same chromosome. There are two possibilities:

- The A and B alleles are on one copy of the chromosome and the a and b alleles are on the other copy. This can be shown like this: $\frac{AB}{ab}$. In this case, the organism will make more AB and ab gametes than aB or Ab.

- The A and b alleles are on one copy of the chromosome and the a and B alleles are on the other copy. This can be shown like this: $\frac{Ab}{aB}$. In this case, the organism will make more Ab and aB gametes than AB or ab.

Work through the diagnostic question on your own and then look at our approach to solving it. If any of the underlined terms are unfamiliar, please consult your book's chapter on linkage and recombination.

Diagnostic Question:

Consider two traits in a hypothetical mammal: size and color. Each trait is controlled by one gene with two alleles. The genes for these two traits are located on the same chromosome and are <u>linked</u>. Large is dominant to small and black is dominant to white.

a) You cross a true-breeding large and black individual with a true-breeding small and white individual. What is the genotype of the resulting offspring (F_1 generation)? Use the notation scheme we showed above.

b) You perform a <u>test cross</u> with an F_1 from part (a) with a small and white individual and obtain the following progeny:

46 Large and black

43 Small and white

5 Large and white

6 Small and black

What is the distance in <u>map units</u> between the size and color genes?

Answer to Diagnostic Question:

a) In the <u>parental generation</u> (P generation), you crossed a true-breeding large and black individual (LLBB) with a true-breeding small and white individual (llbb). We have been told that these two <u>loci</u> are linked so the genotype can be shown as: $\frac{LB}{LB}$ and $\frac{lb}{lb}$. The F_1 offspring will get one chromosome from each parent and will be $\frac{LB}{lb}$.

b) If we were asked to calculate a <u>recombination frequency</u> for these two genes, we could make the calculation:

$$\frac{\text{recombination}}{\text{frequency}} = \frac{\text{\# recombinant progeny}}{\text{total progeny}}$$

In this case, the test cross was $\frac{LB}{lb} \times \frac{lb}{lb}$. The animals that are large and black, $\frac{LB}{lb}$, or small and white, $\frac{lb}{lb}$, are the parental or nonrecombinant types. The nonparental or recombinant types are the large white animals, $\frac{Lb}{lb}$, or the small black animals, $\frac{lB}{lb}$.

Thus, the \# recombinant progeny = 5 + 6 = 11 and the \# total progeny = 46 + 43 + 5 + 6 = 100. Recombination frequency = 11/100 = 0.11 = 11%.
Map units = recombination frequency $\times$ 100. So the size and color genes are 11 map units apart.

Problems

(2.2.1) Consider two traits in a hypothetical mammal: fur color and ear length. Each trait is controlled by one gene with two alleles. The genes for these two traits are located on the same chromosome and are linked. Black fur is dominant to red fur and long ears are dominant to short ears.

a) For your parental generation (P generation), you cross a true-breeding black, long-eared individual with a true-breeding red, short-eared individual. What is the genotype of the resulting offspring (F_1 generation)? Use the notation scheme we showed above.

b) Assume that the genes for these two traits are located on the same chromosome and recombination between the two genes does NOT occur. What types of gametes are produced by the F_1 and in what frequencies will they be produced?

c) Assume that the genes for these two traits are located on the same chromosome and the recombination frequency between the two genes is 10%. What types of gametes are produced by the F_1 from part (a) and in what frequencies will they be produced?

d) Assume that the genes for these two traits are linked with a recombination frequency of 10%. You crossed an F_1 from part (a) with a red, short-eared individual and obtained 100 progeny. Roughly how many offspring of each phenotype would you expect?

e) Suppose you cross a true-breeding black, short-eared individual with a true-breeding red, long-eared individual; this is the parental or P generation. What is the genotype of the resulting offspring (F$_1$ generation)? Use the notation scheme we showed above.

f) Assume that the recombination frequency between the two genes is 10%. What are the types of gametes produced by the F$_1$ from part (e), and in what frequencies will they be produced?

g) Suppose you crossed the F$_1$ from part (e) with a red, short-eared individual and obtained 100 progeny. Roughly how many offspring of each phenotype would you expect?

(2.2.2) You are studying coat color and number of toes in mice. You perform the following crosses:

Cross 1:   Black, five-toed ✗ white, six-toed
                    gives:
             All black, six-toed progeny

Cross 2:   White, five-toed ✗ black, six-toed
                    gives:
             All black, six-toed progeny

a) For each trait (coat color and number of toes), state which is the dominant phenotype and which is the recessive phenotype. Please define your genotype symbols clearly.

b) Assuming that the genes are unlinked, what would you expect if you mated one of the black, six-toed F_1 from cross 2 with a white, six-toed mouse?

c) After performing the cross in part (b), you find:
      3 white, five-toed
      4 black, six-toed
What is one possible explanation for this?

d) You decide to perform several crosses like the cross in part (b) and you find the following:
      57   white, five-toed
      52   black, six-toed
      3   white, six-toed
      5   black, five-toed
What can you conclude from these data?

(2.2.3) You are doing undergraduate research in a genetics lab, working on a hypothetical fly. Your project is to study a few genes in the fly. Preliminary work indicates that the mutant alleles of these genes give recessive phenotypes and the genes are not sex linked, but beyond that you have no information.

You first look at two genes, each with two alleles: "R or r" for body color and "S or s" for wing surface. The red body phenotype is dominant to the yellow body phenotype and smooth wings are dominant to crinkled wings.

You cross a true-breeding yellow-bodied, smooth-winged female with a true-breeding red-bodied, crinkle-winged male.

a) What will be the phenotype(s) of the F_1 progeny?

b) You cross several pairs of F₁ siblings and look at the progeny:

Body	Wing surface	Number
Red	smooth	310
Yellow	smooth	142
Red	crinkled	131
Yellow	crinkled	23

Why don't you see the 9:3:3:1 Mendelian ratio?

c) To determine the recombination frequency between these two genes, you perform several crosses in which you cross an F₁ from part (a) with a yellow-bodied, crinkle-winged fly. You get the following results:

Body	Wing surface	Number
Red	crinkled	396
Red	smooth	102
Yellow	crinkled	98
Yellow	smooth	404

What is the distance between the genes for body color and wing surface in map units?

d) You decide to turn your attention to a different gene, one that controls wing length. This gene has two alleles, "L or l," where long wings are dominant to short wings. Remember that the red body phenotype is dominant to the yellow body phenotype.

You again mate two true-breeding lines:

Red-bodied, short-wing male **X** yellow-bodied, long-wing female
yields
F₁: All red bodied, long-wing

You mate several pairs of F₁ siblings and get the following:

Body	Wing length	Number
Red	long	275
Yellow	long	119
Red	short	124
Yellow	short	8

Why don't you see 9:3:3:1?

e) To determine the recombination frequency between these two genes, you perform several crosses where you cross an F_1 from part (d) with a yellow-bodied, short-winged fly. You get the following results:

Body	Wing length	Number
Red	long	45
Red	short	460
Yellow	long	440
Yellow	short	55

What is the distance between the genes for body color and wing length in map units?

f) Given your answers to parts (c) and (e), what can you say about the relationship between the gene for wing length and the gene for wing surface?

g) Based on the preceding, what are the two possible arrangements of the three genes?

h) In a final mapping experiment, you cross a true-breeding red-bodied, short- and crinkle-winged male with a true-breeding yellow-bodied, long- and smooth-winged female. Diagram the genotype of the F_1 progeny. What is their phenotype?

You cross a male resulting from the cross described in (h) above with a female showing all recessive traits (i.e., yellow with crinkled and short wings, genotype: rrssll) and get:

Phenotype			
Body Color	Wing Length	Wing Surface	Number
Yellow	long	smooth	347
Yellow	short	smooth	45
Yellow	short	crinkled	3
Yellow	long	crinkled	99
Red	short	smooth	91
Red	short	crinkled	356
Red	long	smooth	7
Red	long	crinkled	45

In a cross of this type, the two rarest classes of progeny arise from a double-crossover event. This is shown schematically below:

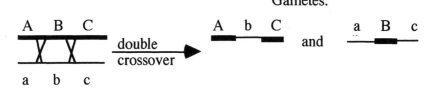

The genotypes of the two least frequent (and therefore double-recombinant) classes of progeny can be used to determine map order by inspection. Only one arrangement of the three genes will give the appropriate genotypes of the double-crossover classes.

i) Using the above data, determine the arrangement of the three genes.

j) For each of the eight classes of progeny listed above, draw the recombination event or lack thereof that occurred in the F_1 male that led to that class of progeny.

(3) CHALLENGE PROBLEMS

All different types of problems can be found in this section. The first part of the challenge is to establish what type of question you are dealing with. Ask yourself: Is the problem concerned with one gene or more than one gene? Does the inheritance pattern fit simple dominance or is it more complicated? Does the question involve linked genes? Once you decide where to begin, these problems may provide additional surprises.

Problems

(3.1) In tomatoes, two alleles of one gene determine the character difference of purple versus green stems. Two alleles of another gene (which segregates independently) determine the leaf shape, sharp versus rounded leaves. Below is a table listing several crosses and the number and type of progeny from each cross.

	Cross	Parental phenotype	Progeny Phenotypes			
			Purple sharp	Purple rounded	Green sharp	Green rounded
a)	1	Purple, sharp ✕ green, sharp	321	101	310	107
b)	2	Purple, sharp ✕ purple, rounded	219	207	64	71
c)	3	Purple, sharp ✕ green, sharp	722	231	0	0
d)	4	Purple, sharp ✕ green, rounded	404	0	387	0
e)	5	Purple, rounded ✕ green, sharp	70	91	86	77

Determine which allele for each locus has the dominant phenotype and which has the recessive phenotype and the most probable genotype for the parents in each cross. Define your symbols clearly.

(3.2) You are studying two traits in fruit flies: wing shape and eye color. Wings can be either bent or straight. Eyes can be either red or yellow. Straight-winged = straight; yellow-eyed = yellow; bent-winged = bent; and red-eyed = red.

Straight-winged, yellow-eyed female ✕ bent-winged, red-eyed male gave:

41 bent yellow males
38 bent red females
38 straight yellow males
43 straight red females

an F_1 bent-winged, red-eyed female ✕ an F_1 bent-winged, yellow-eyed male gave:

31 bent red females	10 straight red females
28 bent yellow females	11 straight yellow females
32 bent red males	9 straight red males
29 bent yellow males	10 straight yellow males

Explain these results and give the genotypes of the parents in each of the crosses.

(3.3) You are interested in two plant genes: one for plant height and one for flower color. You cross a tall white plant to a short red plant. All the progeny are tall pink plants.

a) What is the genotype of the tall white parental plant?

b) What is the genotype of the short red parental plant?

c) What is the genotype of an F_1 tall pink plant?

d) Give six phenotypic classes of plant and the genotype(s) associated with each.

e) Of these six phenotypic classes, which are pure-breeding?

f) If the two genes are unlinked, what offspring do you expect if two tall pink F_1 plants are crossed?

g) If the two genes are linked and no recombination occurs, what offspring do you expect if two tall pink F_1 plants are crossed?

(3.4) In fruit flies, there is a gene called "transformer" that influences sex determination. This problem uses this idea in a hypothetical context.

In the imaginary fruit fly *Drosophila fakeii*, the transformer gene has two alleles: the T allele allows for normal sex determination (XX = female; XY = male), but the t allele correlates with transformed sex determination. This allele transforms genetic males into females, but has no effect on genetically female flies (XX = female; XY = female). This leads to the following:

Genotype	Phenotype
XX T_	female
XX tt	female
XY T_	male
XY tt	female

Eye color in this species is controlled by an X-linked gene with two alleles, X^R and X^r, where red eyes are dominant to white.

a) The following cross is performed:

   tt $X^R X^r$ red female ✕ Tt X^RY red male

Predict the phenotypic classes of progeny and explain why this results in a 3:1 ratio of females/males.

b) What phenotypic ratios would be seen in the progeny of the cross described in part (a) if the phenotype of the transformer gene was recessive male lethal (this is also observed in real flies)? That is:

Genotype	Phenotype
TT or Tt	live
tt	male lethal (XX = female; XY = dead)

This leads to the following:

Genotype	Phenotype
XX T_	female
XX tt	female
XY T_	male
XY tt	dead (not seen in offspring at all)

(3.5) Cystic fibrosis is an autosomally inherited genetic disease. The disease phenotype is recessive to the normal phenotype.

a) If two normal individuals who are carriers (genotypically heterozygous for the disease allele but phenotypically normal) have a child, what is the probability that the child will have cystic fibrosis?

b) If two normal phenotype parents (genotype unknown) have a child with cystic fibrosis, what are the chances that their second child will have cystic fibrosis?

c) If the frequency of carriers in the human population is 0.001 and an individual with cystic fibrosis has a child with an individual who is phenotypically normal (genotype unknown), what is the probability that the child will have cystic fibrosis?

(3.6) Consider an autosomal recessive genetic disease, this time a hypothetical one called Qosis that is caused by the absence of the enzyme Qase. You would like to have some method that would allow you to test couples to determine whether the children they will have will be at risk of having Qosis.

However, there is no simple enzymatic assay (test) for Qase. In this hypothetical case, you are fortunate because the gene for Qase is linked to the gene for blood type. That is, the Qase gene (with two alleles: normal and Qosis) lies on the same chromosome and very close to the gene for blood type (with three alleles: I^A, I^B, and i).

For the purposes of this question, you should assume:
     1) No recombination occurs between the Qase gene and the blood type gene.
     2) No new mutations in the Qase gene occur over the time of this pedigree.
     3) No one who marries into this family is a carrier of Qosis.

Given the pedigree, answer the following questions. The blood type of each individual is indicated.

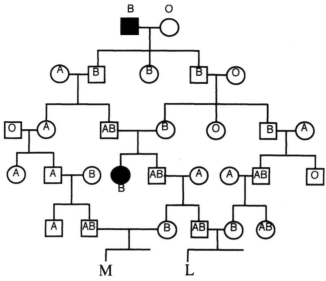

a) Which blood-type allele is the Qosis allele linked to in this family? Explain briefly how you arrived at this conclusion.

b) What is the risk that child L has Qosis if his/her blood type is A? B? AB? Explain briefly.

c) What is the risk that child M has Qosis if his/her blood type is A? B? AB? Explain briefly.

d) If there were some recombination between the gene for Qase and the gene for blood type, would this complicate your analysis? Explain briefly.

(3.7) This problem requires that you extend your knowledge of recombination from the theoretical to a more timely situation: genetic testing. It is designed to give you practice understanding chromosomes and linkage in a new context.

The fictional mopple is an extremely rare diploid species that mates for life and has few offspring (cubs). The biologists who study this species are concerned about a recessive disease that is shortening the animals' lives. They call upon you, a world-renowned geneticist, to help out. Since it is inappropriate to do invasive sampling and this population has been monitored for many years, you do a pedigree analysis of this species. You know that the gene for eye color is linked to the disease gene. You construct the following pedigree for one family of mopples, which shows the phenotypes for eye color (B, blue; R, red) and the disease (shaded individuals have the disease). Eye color is controlled by one gene with two alleles. Red eyes are dominant to blue eyes.

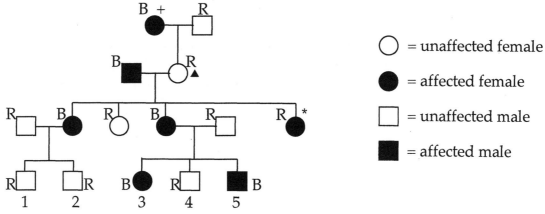

○ = unaffected female

● = affected female

□ = unaffected male

■ = affected male

a) What are the genotypes for the individuals numbered 1–5? Use the symbols:

Genotype	Phenotype
RR or Rr	red eyes
rr	blue eyes

Genotype	Phenotype
D^+D^+ or D^+D^-	no disease
D^-D^-	diseased

Note: since the two genes are linked, it will be most convenient to use notation similar to this (shown for the individual indicated by a [+]):

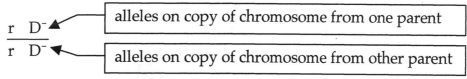

r D^- ◄ — alleles on copy of chromosome from one parent

r D^- ◄ — alleles on copy of chromosome from other parent

b) You know that the genes for eye color and the disease are linked. To which eye color allele is the disease allele linked in this family?

c) What must the genotype of the affected red-eyed female mopple (indicated by an *) be? Explain how this red-eyed affected mopple arose. Include a schematic of the event in your explanation.

d) Suppose you knew that the distance between the eye color and the disease loci was 15 map units (15% recombination). If the parents of the (*) mopple had a **sixth** cub with red eyes, what is the risk (given as a percent chance) that this cub could be affected with the disease?

(3.8) You are studying a human disease to see whether it is inherited or not. You construct the following pedigree and find that it is not consistent with any of the usual modes of inheritance.

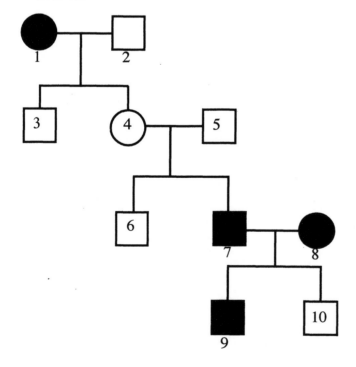

Chapter 1: Genetics Problems

a) Which part of the pedigree is inconsistent with <u>autosomal dominant</u> inheritance? Redraw that portion in the space below and explain why it is **inconsistent** with <u>autosomal dominant</u> inheritance. If more than one part of the pedigree is appropriate, you need draw only one part. You may explain in words or by using genotypes as you prefer.

b) Which part of the pedigree is inconsistent with <u>autosomal recessive</u> inheritance? Redraw that portion in the space below and explain why it is **inconsistent** with <u>autosomal recessive</u> inheritance. If more than one part of the pedigree is appropriate, you need draw only one part. You may explain in words or by using genotypes as you prefer.

c) Which part of the pedigree is inconsistent with <u>sex-linked recessive</u> inheritance? Redraw that portion in the space below and explain why it is **inconsistent** with <u>sex-linked recessive</u> inheritance. If more than one part of the pedigree is appropriate, you need draw only one part. You may explain in words or by using genotypes as you prefer.

d) If you changed **one** of the <u>unaffected</u> individuals to <u>affected</u>, the pedigree would be consistent with <u>autosomal recessive</u> inheritance. Which individual would you change? If more than one is possible, give only one.

(3.9) You are helping Sigourney Weaver study those nasty sci-fi aliens. You have found that the aliens are diploids, their genes behave like those of earth animals, and the two different sexes have different sex chromosomes, but you are unsure whether XX is female and XY is male or ZZ is male and ZW is female.

You discover that they have three blood types, which you label type α, type β, and type γ. While wearing a protective suit, you perform the following crosses and observe the results:

Cross 1:   male α ✕ female β
     progeny: 25% male α
                  25% male β
                  25% female γ
                  25% female β

Cross 2:   male β ✕ female γ
     progeny: 50% male α
                  50% female β

Cross 3:   male γ ✕ female β
     progeny: 50% male α
                  50% female γ

Cross 4:   male β ✕ female β
     progeny: 50% male β
                  50% female β

Cross 5:   male γ ✕ female γ
     progeny: 50% male γ
                  50% female γ

Cross 6:   male α ✕ female γ
     progeny: 25% male α
                  25% male γ
                  25% female γ
                  25% female β

a) Based on these data, construct a simple, plausible model for the inheritance of blood type in these organisms. Be sure to indicate:
- the number of genes and alleles involved
- the genotypes that correspond to each blood type
- whether any of the blood type genes are located on sex chromosomes, and
- which sex determination system is being used; i.e., are females XX or ZW?

b) No females with blood type α were described above. Based on your model, is it possible to find a female with blood type α? Explain your reasoning.

Chapter 2:

Biochemistry Problems

Biochemistry Problems

If you were a biochemist, you would study chemical substances and vital processes that occur in living organisms. You might study macromolecules such as lipids and phospholipids, carbohydrates, proteins, or nucleic acids. You might study pathways such as glycolysis or photosynthesis, or any other metabolic pathway. In this chapter, we begin with problems that review the bonds and forces that hold these macromolecules together. We briefly touch on macromolecules that are not proteins, but the majority of this chapter asks you to explore the structure and function of proteins.

(1) BONDS AND FORCES

(1.1) Covalent bonds

For the purposes of this book, we have simplified the covalent bonding properties of the atoms most commonly found in living organisms. For this book, we will use the bonding properties given in the following chart:

| | Number of covalent bonds | | | | | |
Element	0	1	2	3	4	5
H	+	neutral				
O		–	neutral			
N				neutral	+	
C					neutral	
S			neutral			
P						neutral

The shaded boxes indicate configurations that do not appear in this book (for example, a sulfur atom making three covalent bonds). These approximations are sufficient for the problems in this book and most introductory biology courses. As you take further courses in biology and chemistry, you will learn about additional possibilities.

Diagnostic Question:

Convert the following shorthand formulas to correct structural formulas.
For example:

$$CH_4 \xrightarrow{\text{becomes}} \begin{array}{c} H \\ | \\ H-C-H \\ | \\ H \end{array}$$

Carbon makes 4 bonds; hydrogen makes 1.

a) H_3CCH_3

b) C_2H_4

c) $H_2N(CH_2)_3CH_3$

d) $(CH_3)_3N^+CH_2CH_2OH$

e) CH_3COOH

Answer to Diagnostic Question:

a)

H3CCH3

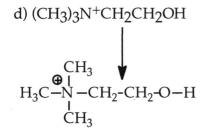

Carbon makes 4 bonds;
hydrogen makes 1.

b) C_2H_4

from the formula:

H   H
  C–C
H   H

but the C's are making
only three bonds
so add a double bond:

H   H
  C=C
H   H

(correct structure)

c) $H_2N(CH_2)_3CH_3$

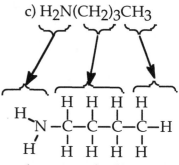

H H H H
H | | | |
 N–C–C–C–C–H
H | | | |
    H H H H

Note that since the (CH2)3
is CH2 and not CH3, the carbons
must be in a line so that the C's
can make 4 bonds.

d) $(CH_3)_3N^+CH_2CH_2OH$

```
        CH3
         ⊕ |
H3C–N –CH2-CH2-O–H
         |
        CH3
```

Nitrogen making 4 bonds has a (+) charge;
oxygen makes 2 bonds.

e) CH_3COOH

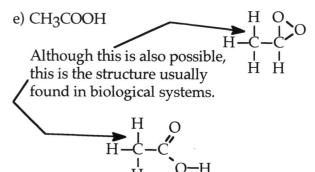

Although this is also possible,
this is the structure usually
found in biological systems.

Problems:

(1.1.1) Check the following structures and correct any mistakes you find. There may be
more than one way to correct the structure.

(1.1.2) For each of the functional groups given, draw a structural formula.

- Amino

- Hydroxyl

- Carboxyl

- Methyl

- Phosphoryl

- Aldehyde

(C1) Computer-Aided Problems 1

These problems use the Molecular Calculator that allows you to practice drawing structural formulas and working with simplified structures. The program, "Molecular Calculator," is on the CD in the "Biochemistry" folder. You can access it by double-clicking on its icon.

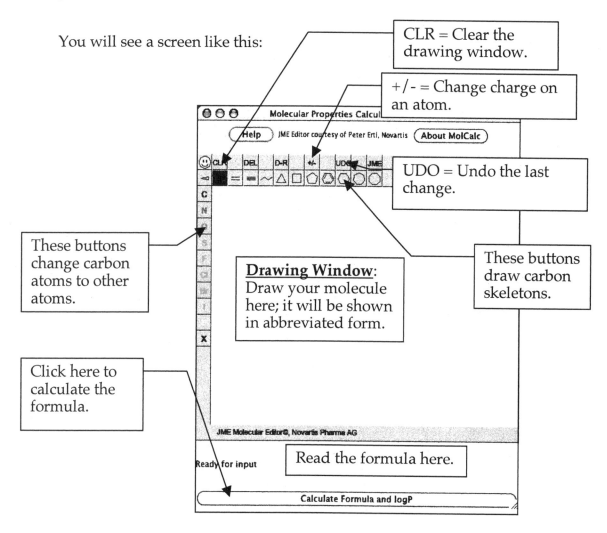

You will see a screen like this:

CLR = Clear the drawing window.

+/- = Change charge on an atom.

UDO = Undo the last change.

These buttons change carbon atoms to other atoms.

Drawing Window: Draw your molecule here; it will be shown in abbreviated form.

These buttons draw carbon skeletons.

Click here to calculate the formula.

Read the formula here.

This program will let you draw molecules that follow the basic rules for covalent bonding shown in the table below.

	Number of covalent bonds					
Element	0	1	2	3	4	5
H	+	neutral				
O		–	neutral			
N				neutral	+	
C					neutral	
S			neutral			
P						neutral

This first practice exercise is designed to familiarize you with the Molecular Calculator. You will build a molecule and calculate its formula. The molecule is shown below:

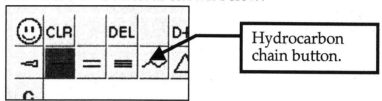

The following steps show you how to draw this molecule and give you practice with the software.

1) Draw propane ($H_3CCH_2CH_3$ or ‿). To do this, you:
   a) Click on the "hydrocarbon chain" button as shown below:

Hydrocarbon chain button.

   b) Put the cursor in the middle of the screen and drag quickly to the right. You will see a zigzag line forming near the cursor, and a number will appear in the lower left part of the Drawing Window. The line is the chain of carbons, and the number tells you how many carbons long it is. Stop when you get to 3. When you release the mouse, you should see something like this:

   If you make a mistake, you can either:
   • Clear it all by clicking the "UDO" (Undo) button at the top of the Drawing Window.
   • Click the "DEL" (delete) button at the top of the Drawing Window. This will delete whatever you click on.

If you want to move your molecule, click near the molecule but not on an atom, and drag the molecule to a more convenient place.

2) Calculate the formula of propane. Click the "Calculate Formula and logP" button; we'll deal with "logP" in a later problem. The calculation may take a few seconds. You should see something like this:

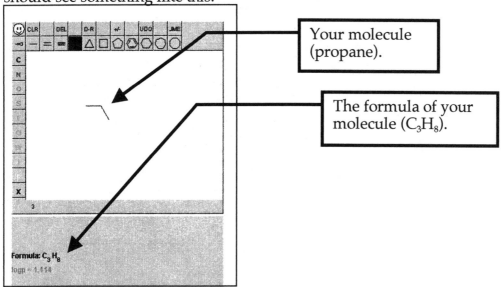

Your molecule (propane).

The formula of your molecule (C_3H_8).

Propane's formula should be C_3H_8.

3) Change propane to phenyl propane by adding a benzene ring.
    a) First, add a single bond from the left-most carbon using the "bond" tool.
    b) Click on the "bond" tool until it turns dark gray.
    c) Move the cursor over the left-most carbon until you see a blue square appear.

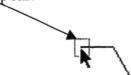

    d) Click once to add a carbon and you should see:

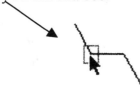

    e) Now add a benzene ring with the benzene ring tool. First, click on the benzene ring tool:

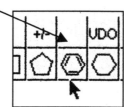

f) Move the cursor until a blue square appears at the left-most carbon in the chain you made.

g) Click the mouse and you should see:

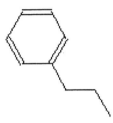

4) Calculate the formula of phenyl propane as you did for propane (step 2). The formula should be C_9H_{12}.

5) Change the molecule one last time.

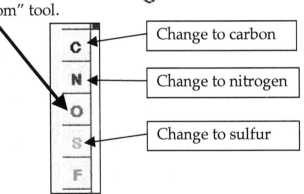

a) Use the "bond" tool as you did in step 3 (a) through (d) to add a carbon to the second carbon from the right-hand end of the chain. Your molecule should look like this:

b) Select the "Change to Oxygen atom" tool.

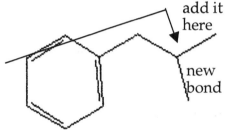

c) Move the cursor to the end of one of the branches at the end of the chain until you see a light-blue square appear.

d) Click on the atom to change the carbon to oxygen. You should see:

e) Do the same at the other branch end and you should see:

f) Change one of the OH's to O⁻. Click on the "+/-" tool and click on one of the OH's. You should see:

g) To make the structure complete, you must make a double bond between the O (not the O⁻) and the carbon. Do this by selecting the "bond" tool and moving it over the <u>bond</u> between the O and the carbon until you see a blue <u>rectangle</u> appear. Click once to make it a double bond. You should see:

6) Calculate the formula of your new molecule as you did for propane (step 2). The formula should be $C_8H_7O_2$ (–).

7) Draw several molecules on the screen and calculate their formulas by hand. Check your work by clicking the "Calculate Formula and logP" button.

Note that it is possible to draw molecules that the software cannot process properly. Some of these molecules are chemically possible, but their chemistry is beyond the scope of this book. If you attempt to calculate the logP and formula of a molecule containing any of the following atoms, the program will tell you that "It is not possible to calculate logp...". These "illegal" atoms are:
- A carbon atom with any charge.
- A (+)-charged S or O atom.
- An N-atom with a (–) charge or with a charge greater than (+1).
- An S-atom making 3 or 5 bonds.
- A charged P-atom or a P-atom making more or less than 5 bonds.
- A charged F, Cl, Br, or I atom.
- An "X" atom.

(1.1.3) For each of the following formulas, draw a molecule with the same formula.
a) C_3H_8O

b) C_3H_6

c) C_3H_5NO

d) $C_2H_4NOS(-)$ (This molecule has a single negative charge on one atom.)

e) $C_5H_8N(+)$ (This molecule has a single positive charge on one atom.)

(C2) Computer-Aided Problems 2

For the next problems, you will use computer software that allows you to manipulate a two-dimensional (2-d) representation of a three-dimensional (3-d) molecule. This software is called molecular visualization (MolVis) software. The MolVis software you will use is called "Molecules in 3-d" and can be found in the "Biochemistry" folder on the CD-ROM that came with this book.

Objectives:
To familiarize you with:
- The structures of some important biomolecules that you will see again and again.
- Translations between the 2-d representations you see in this and other books and the 3-d reality of biomolecules.
- The kind of representation used by the MolVis software that you will use in this book.
- The user interface of the MolVis software that you will use in this book.

Start "Molecules in 3-d" in the "Biochemistry" folder, and click the "maximize" button. On a PC, it is in the upper-right corner of the window and looks like this:
On the Macintosh, it is in the upper-left corner of the window and looks like this (the green button with the + sign):

Note that Molecules in 3-d can sometimes take a little while to get started. You will see the following on a PC:

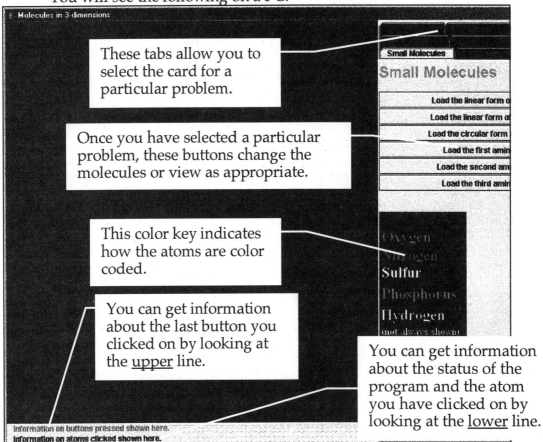

On a Macintosh, the problem selection tab looks like this (the rest is very similar to the screen shown above). You can see the tabs for two problems; you access the remainder by clicking the arrow.

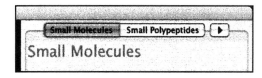

We will use this software throughout this part of the book, so we will take some time now to describe its use in detail. This software allows you to get information from the image in several ways:

- Rotating the molecule: This is the best way to get an idea of the molecule's 3-d structure. You can click and drag on any part of the molecule and it will rotate as though you had grabbed it.
- Zooming in or out: This helps to get close-up or "big-picture" views of the molecule. Hold the shift key down while dragging the cursor up (to zoom out) or down (to zoom in) the image.
- Identifying the atom you are looking at: You can find information on the atoms in the molecule in one of two ways:
  - By clicking on an atom and looking at the lower left of the Molecules in 3-d window. A small line of text will appear there with information on the atom you just clicked.
  - By putting the cursor over the atom you are interested in and waiting a few seconds for the information to pop up. The program will then display information on the atom in a little pop-up window. The information in the pop-up is more detailed than the first one above but rather cryptic. If you put the cursor over the left-most carbon atom (gray), the pop-up reads "1.C. #7." The most important part of this is the "C"; this says that you clicked on a carbon atom. Try putting the cursor over some other atoms to see what you get. Note that this does not always work, especially on Macintosh computers.

In addition to the above, atoms are also identified by their color. The color scheme is shown to the right of the molecule images.

Atoms are indicated by spheres; covalent bonds are shown by rods; noncovalent bonds are not shown at all.

Important note: This software does not distinguish between single, double, and triple covalent bonds. All covalent bonds are shown as single rods. You have to decide whether a bond is single, double, or triple based on your knowledge of covalent bonding and the structures of known biological molecules.

Click the tab for this problem "Small Molecules." Click the button marked "Load the linear form of glucose" and you should see this in the large black window (the molecule window):

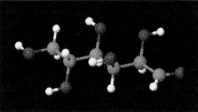

This is a 2-d representation of glucose. Since glucose is really 3-dimensional, you can't see all the details of its structure from a single 2-d image.

Each of the following questions applies to the structures shown by the program.

a) Click the button marked "Load the linear form of glucose." Note that the top line of text below the structure now shows "Load the linear form of glucose"; this is to remind you which structure you are looking at. The structure shown is the sugar glucose in its linear form. Based on the image shown, draw the structure of the linear form of glucose. You should use letters to represent atoms and lines to represent covalent bonds. Be sure to include all hydrogens. Compare this structure with the structure of glucose given in your textbook.

b) Click the button marked "Load the linear form of fructose." This shows the sugar fructose in its linear form. Based on the image shown, draw the structure of the linear form of fructose. Compare this structure with the structure of fructose given in your textbook. How does fructose differ from glucose?

c) Click the button marked "Load the circular form of glucose." This shows the structure of glucose in its circular form. Based on the image shown, draw the structure of the circular form of glucose. Which parts of the linear glucose molecule were connected to give the circular form? Hint: it involves attaching one atom to another and moving one hydrogen atom; no carbon-carbon bonds are made or broken.

d) Click the button marked "Load the first amino acid." This shows an amino acid. Draw its structure and determine which amino acid it is.

e) Click the button marked "Load the second amino acid." This shows an amino acid. Draw its structure and determine which amino acid it is.

f) Click the button marked "Load the third amino acid." This shows an amino acid. Draw its structure and determine which amino acid it is.

(C3) Computer-Aided Problems 3

In these problems, you will use the same molecular visualization software you used in problem (C2) to determine the structure of two short proteins. Each of these consists of three amino acids linked by peptide bonds; they are therefore called "tripeptides."

Objectives:

To familiarize you with:

- The way that the MolVis software displays protein structures – specifically that hydrogen atoms are not shown.
- The structure of amino acids in proteins – specifically, the backbone, side chains, and peptide bonds.
- The chemical meaning of the terms amino and carboxyl terminus as well as the directionality of the protein sequence.

You can access this problem by launching "Molecules in 3-d" in the "Biochemistry" folder on the CD-ROM and selecting the tab for this problem "Small Polypeptides." As a reminder, here are a few important notes about the way these molecules are displayed:

- The atoms are colored according to the scheme at the right side of the window.
- Only covalent bonds are shown.
- All covalent bonds are shown as single lines; single, double, and triple bonds are all shown identically. You have to figure out the bond type based on your knowledge of covalent bonding and amino acid structures.
- In contrast to the structures in problem (C2), **hydrogen atoms are not shown** in these structures. This is because these are actual protein structures determined by X-ray crystallography. Hydrogen atoms are not visible in X-ray crystallograms and are therefore not shown in these structures. You have to place the hydrogens yourself based on your knowledge of covalent bonds and amino acid structures.

Some hints:
- Consult your textbook for the structure of protein molecules. Be sure to note how to identify the backbone, side chains, and peptide bonds.
- Trace the backbone first. Click on atoms to identify them: the backbone atoms are labeled "N" for the amino nitrogen; "CA" for the alpha-carbon (the carbon that the side chain is attached to); and "C" for the carboxyl carbon.
- Draw the skeleton structures based only on what you can see in the web page first. You can fill in the details (double bonds, hydrogen atoms, and charges) later.
- Note that there may be some ambiguities, for example:
  - A nitrogen atom connected by one line ("-N" in the skeleton structure) can be either $-NH_3^+$ or $-NH_2$.
  - An oxygen atom connected by one line can be either -OH or $-O^-$.
  If you cannot tell from the information given, then either structure is OK.
- Use a table of amino acid structures to help you trace and draw the complete structure. Remember that these have to be proper amino acids; not just any structure is possible.

a) Click the button marked "Load the first tripeptide." Based on the structure shown, answer the following questions:

   i) Draw the structure of the molecule shown; be sure to include all hydrogen atoms.

ii) Indicate each of the two peptide bonds in the tripeptide with an asterisk (*).

iii) Write out a shorthand representation of the protein structure in the following format:

N-aa1-aa2-aa3-C

Where N is the amino terminus; aa1 is the name of the amino-terminal amino acid; aa2 is the name of the middle amino acid; aa3 is the name of the carboxyl-terminal amino acid; and C is the carboxyl terminus. For example: "N-alanine-glycine-threonine-C."

b) Click the button marked "Load the second tripeptide." Based on the structure shown, answer the following questions:

i) Draw the structure of the molecule shown; be sure to include all hydrogen atoms.

ii) Indicate each of the two peptide bonds in the tripeptide with an asterisk (*).

iii) Write out a shorthand representation of the protein structure in the following format:

N-aa1-aa2-aa3-C

Where N is the amino terminus; aa1 is the name of the amino-terminal amino acid; aa2 is the name of the middle amino acid; aa3 is the name of the carboxyl-terminal amino acid; and C is the carboxyl terminus. For example: "N-alanine-glycine-threonine-C."

(1.2) Noncovalent bonds and forces

In these problems, you will be given the covalent bonds (these are shown as solid lines) and must *infer* their noncovalent bonding properties. Noncovalent bonds/interactions are shown by dotted lines (etc.). These two types of "bonds" are entirely separate; for example, an oxygen (which can make only two covalent bonds) can make several hydrogen bonds *in addition to* the covalent bonds. That is, noncovalent "bonds" do not count toward an atom's covalent bond total.

As a reminder, here are the types of noncovalent interactions we will use in this book. They are listed from strongest to weakest:

- Ionic/electrostatic bonds (also known as "salt bridges"): These are the strongest noncovalent bonds. They occur between *fully charged* (that is, + or –; not partially charged) atoms or groups.
- Hydrogen bonds: These require a "hydrogen donor": a hydrogen atom covalently bonded to an oxygen or nitrogen (–OH or –NH) and a "hydrogen acceptor": a lone pair of electrons on an oxygen or nitrogen atom (O: or N:).
- Hydrophobic interactions: These occur when several or many hydrophobic atoms or groups clump together to avoid contact with water. Hydrophilic groups cannot form hydrophobic interactions. Unlike an ionic or a hydrogen bond that occurs between two molecules, hydrophobic interactions are not true bonds, but involve nonpolar molecules that cluster together to avoid the water that surrounds them. The effect of clustering nonpolar molecules or chemical groups to shield them from water is a significant force.
- van der Waals bonds: These occur between any two nonbonded atoms and are the weakest interactions possible. Although they are always present, they are not significant unless large surface areas are positioned very closely together. In this case, the combined van der Waals forces can play a significant role.

There are several ways you will be asked to apply this information depending on the nature of the question.

First, the question can be asked in one of two ways:

1) What types of interactions are *possible*? In this case, there can be more than one answer. You should specify all the types that could occur.

2) What is the *strongest* interaction between two molecules? In this case, we assume that the strongest noncovalent interaction is an ionic bond, followed by a hydrogen bond, and finally a van der Waals interaction. If there are several nonpolar interacting species, then hydrophobic interactions should be considered.

Second, the question can be asked in one of two contexts:

1) <u>What kind(s) of interaction(s) can this part of a molecule make?</u> Since it takes two items to make a bond, the bond couldn't form without a "suitable partner." Either explicitly or implicitly, this question assumes the existence of a suitable partner. For these, the following flowchart applies:

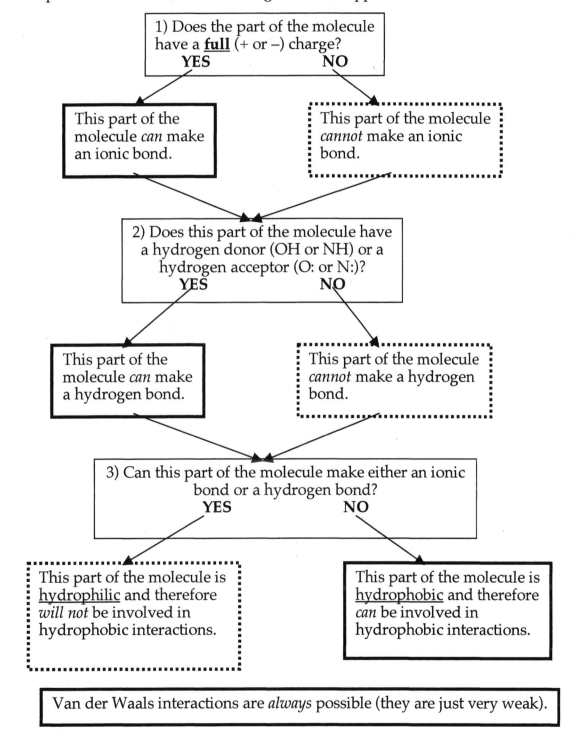

2) <u>What kind(s) of interactions are possible between these two (parts of) molecules?</u> In this case, you have to determine whether the other molecule is a suitable partner. This is a slightly more restrictive question than (1). The flowchart below applies in this case. Note that the questions now ask about the other molecule(s).

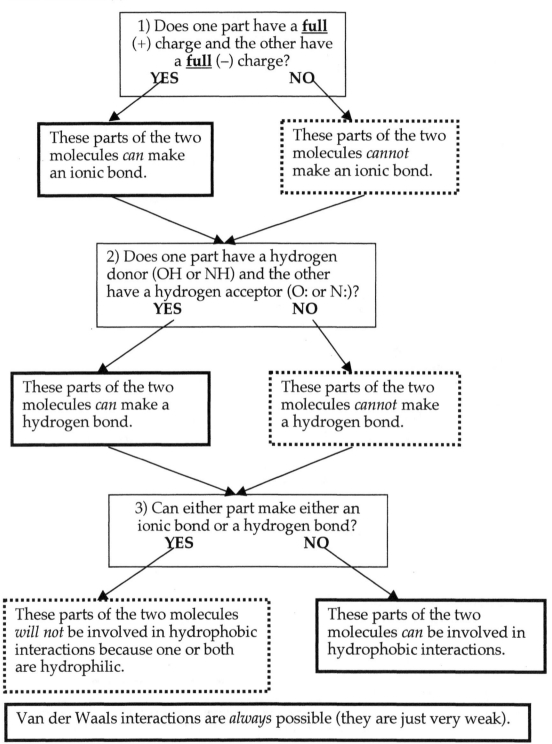

Chapter 2: Biochemistry Problems

You will also be asked to compare the relative hydrophobicity/hydrophilicity of different molecules. For these problems, the following rules are useful:

1. The more hydrophilic atoms or groups of atoms that a molecule has, the more hydrophilic the molecule is. Hydrophilic groups are:
   - Charged (+) or (–)
   - Hydrogen bond donors (NH or OH)
   - Hydrogen bond acceptors (N: or O:)

2. Charged groups are more hydrophilic than hydrogen bond donors or acceptors.

3. The more hydrophobic atoms or groups that a molecule has, the more hydrophobic the molecule is. Hydrophobic groups are any not listed above (for example, C–H, S–H, C–C, C–S, C=C, etc.).

4. Per atom or group of atoms, hydrophilic groups contribute more than hydrophobic groups to the overall hydrophobicity of a molecule. That is, one hydrophilic group will make a molecule more hydrophilic than one hydrophobic group will make it hydrophobic. Put another way, imagine putting parts of a molecule on a scale with hydrophobic parts on one side and hydrophilic parts on the other. Each of the hydrophilic groups will "weigh" more than each of the hydrophobic groups. Thus, it takes more hydrophobic atoms to "balance out" a single hydrophilic atom.

Diagnostic Question:

Complete the table below. When evaluating the bond or interaction, assume that a suitable partner is nearby.

Part of molecule	Is the bond polar or nonpolar?	Hydrophobic or hydrophilic?	Ionic bond?	Hydrogen bond?	Hydrophobic interactions?
C–C					
C–H					
C–N					
C–O					
S–H					
O–H					
N–H					

Answer to Diagnostic Question:

Part of molecule	Is the bond polar or nonpolar?	Hydrophobic or hydrophilic?	Ionic bond?	Hydrogen bond?	Hydrophobic interactions?
C–C	*nonpolar*	*hydrophobic*	*no*	*no*	*yes*
C–H	*nonpolar*	*hydrophobic*	*no*	*no*	*yes*
C–N	*polar*	*hydrophilic*	*a*	*‡*	*no*
C–O	*polar*	*hydrophilic*	*a*	*‡*	*no*
S–H	*nonpolar*	*hydrophobic*	*no*	*no*	*yes*
O–H	*polar*	*hydrophilic*	*a*	*yes*	*no*
N–H	*polar*	*hydrophilic*	*a*	*yes*	*no*

[a] If the O or N is charged, "yes"; if not, "no."
‡ Yes, if the N or O has a lone pair available.

Problems:

(1.2.1) Complete the table below. When evaluating the bond or interaction, assume that a suitable partner is nearby.

Part of molecule	Is the bond polar or nonpolar?	Hydrophobic or hydrophilic?	Ionic bond?	Hydrogen bond?	Hydrophobic interactions?
C–S					
P–O					
S–O					
—N—					
—N⊕—					
—O—					
—O⊖					
—S—					

(1.2.2) A gecko can stick to just about any surface and walk with its feet over its head. The sole of a gecko's foot is covered with perhaps a billion tiny hairs that put the gecko in direct physical contact with its environment. In experiments, the toes of geckos adhered equally well to neutral, strongly hydrophobic, and strongly hydrophilic surfaces. As the number of tiny hairs decreases, the adhesive properties decrease. What noncovalent force or bond might explain the gecko's acrobatics?

(1.2.3) For each molecule, draw a solid line around each hydrophilic group of atoms; draw a dotted line around each hydrophobic group of atoms. For each group you circle, give the type(s) of bonds that this group could make (ionic bond, hydrogen bond, hydrophobic interaction).

For example:
aspirin

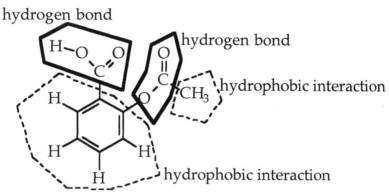

a) Soap

$$\overset{\ominus}{O}\overset{O}{\underset{}{\overset{\|}{C}}}-CH_2-CH_2-CH_2-CH_2-CH_2-CH_2-CH_2-CH_2-CH_2-CH_2-CH_2-CH_2-CH_2-CH_2-CH_2-CH_3$$

b) Phenylalanine (an amino acid)

The hydrogens are often off of the ring for simplicity.

(1.2.4) Draw the hydrogen bonds that could form between water molecules and the appropriate regions of arginine. Indicate the hydrogen bonds with dashed lines.

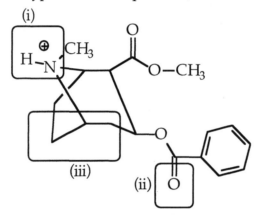

(1.2.5) Shown below is the structure of cocaine. For each of the circled regions, indicate which bonds that part of cocaine could form with another molecule, given a suitable partner. Assume that the circled parts remain attached to the rest of the molecule. Fill in the table with "yes" if that type of bond is possible, "no" if it is not.

Part	Could this part form **ionic bonds** with another molecule?	Could this part form **hydrogen bonds** with another molecule?	Could this part form a **hydrophobic interaction** with another molecule?
(i)			
(ii)			
(iii)			

(1.2.6)
Rank these in order from most hydrophobic to most hydrophilic and explain.

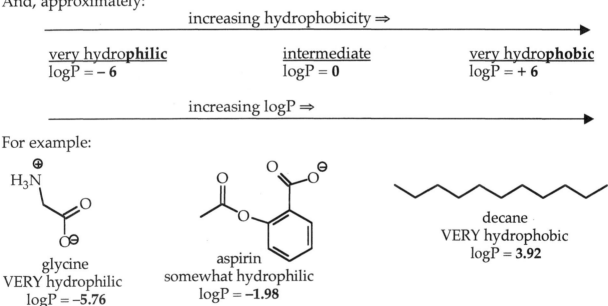

(C4) Computer-Aided Problems 4
The Molecular Calculator is a computer program that calculates the relative hydrophobicity of a molecule. The program calculates the hydrophobicity of a molecule in terms of its logP (short for "log $P_{O/W}$"). You will draw molecules and the program will calculate the approximate hydrophobicity of the molecule.

The value of logP tells you how hydrophobic a molecule is. For more detail, see the end of this problem. The higher the logP value, the more hydro**phobic** the molecule is. And, approximately:

increasing hydrophobicity ⇒

very hydro**philic**   intermediate   very hydro**phobic**
logP = – 6   logP = 0   logP = + 6

increasing logP ⇒

For example:

glycine
VERY hydrophilic
logP = –5.76

aspirin
somewhat hydrophilic
logP = –1.98

decane
VERY hydrophobic
logP = 3.92

You will use the Molecular Calculator to check your own estimations of relative hydrophobicity as a way to practice with this material.

You will use the Molecular Calculator as you did in problem (C1) to work through the following problems. To calculate the logP value of a molecule, click "Calculate Formula and logP." Look at the "logP" value shown at the bottom of the window.

1) Consider the following three molecules:

Molecule #1   Molecule #2   Molecule #3

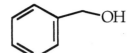

a) Rank them in order from most hydrophobic to least hydrophobic using what you know about chemical properties. Explain your choices.

Most hydrophobic   Intermediate   Most hydrophilic

b) Use the Molecular Calculator to check your answer.

Molecule	logP
1	
2	
3	

c) Make a molecule more hydrophobic than the most hydrophobic molecule from part (1a). Check your work with the Molecular Calculator.

logP:_____

d) Make a molecule more hydrophilic than the most hydrophilic molecule from part (1a). Check your work with the Molecular Calculator.

logP:_____

e) Make a molecule that is in between two of the molecules from part (1a) in terms of hydrophobicity. Check your work with the Molecular Calculator.

logP:_____

2) Different groups of atoms contribute differently to the logP of a molecule. This question compares the contributions of four different groups of atoms. In organic chemistry "R" is shorthand used to represent "the rest of the molecule." To answer this question, you can use the "R group" of your choice; just be sure that you use the same "R group" for all four molecules.

Consider the following four molecules:

$R\text{-}CH_3$ $R\text{-}OH$ $R\text{-}SH$ $R\text{-}NH_2$

For any given R group, two have high logP values and two have low logP values.

a) Choose an R group of your own design, draw the four variations of this molecule ($R\text{-}CH_3$, $R\text{-}OH$, $R\text{-}SH$, and $R\text{-}NH_2$), and give their logP values. Note that you can calculate the formula to be sure that you have done this correctly. Suppose that you started with a particular R group. If you add a $-CH_3$, one of the H's will be replaced by a CH_3; so the new formula should be "R" minus one H (for the one that was replaced) plus one C and three H's. Overall, this would be "$R + C + H_2$." Likewise for $R\text{-}OH$, the new formula should be "$R + O$"; for $R\text{-}SH$, "$R + S$"; and for $R\text{-}NH_2$, "$R + N + H$."

b) In terms of the polarity of the bonds involved, explain why the two molecules with high logP are more hydrophobic and why the two with low logP are more hydrophilic.

3) Ethanol (H_3CCH_2OH) and di-methyl-ether (H_3COCH_3) have the same number of carbons, hydrogens, and oxygens (C_2H_6O) but differ in the following important way. In ethanol, the O is bonded to a carbon and a hydrogen, but in di-methyl-ether, the O is bonded to two carbons.

Create a similar pair of molecules; you can check these features by having the program calculate the formula for you.
- Both members of this pair should have the same number of carbons, hydrogens, and oxygens.
- Both members should have only one oxygen.
- One member should have the oxygen bonded to a carbon and a hydrogen; the other should have the oxygen bonded to two <u>different</u> carbon atoms.

a) Draw the two molecules.

b) In terms of their capability of forming bonds with water, predict which will be more hydrophobic and explain your reasoning.

c) Give the logP values for your two molecules. Do they agree with your prediction?

4) Adding an -OH (hydroxyl) group makes a molecule more hydrophilic; adding a -CH$_3$ (methyl) makes a molecule more hydrophobic. Approximately how many -CH$_3$'s are required to counterbalance the effect of an -OH? Note that this will depend on many factors and will not be the same for all molecules.
a) Start with a molecule of your choosing. Draw it below and calculate its logP:_____

b) Add an -OH to the molecule from part (4a). Draw it below and calculate its logP:_____

c) Keep adding -CH$_3$'s to the molecule from part (4b) until it has approximately the same logP as the original molecule (4a). Draw the molecule below, fill in the number of -CH$_3$'s you had to add, and give the logP.

of -CH$_3$'s required:_____

logP:_____

5) Adding a charged group $-O^-$ or $-NH_3^+$ group makes a molecule <u>much</u> more hydro<u>philic</u>; adding a $-CH_3$ (methyl) makes a molecule more hydro<u>phobic</u>. Approximately how many $-CH_3$'s are required to counterbalance the effect of a charged group? <u>Note</u> that this will depend on many factors and will not be the same for all molecules.

a) Start with a molecule of your choosing. Draw it below and calculate its logP:_____

b) Add a charged group to the molecule from part (5a). Draw it below and calculate its logP:_____

c) Keep adding $-CH_3$'s to the molecule from part (5b) until it has approximately the same logP as the original molecule (5a). Draw the molecule below, fill in the number of $-CH_3$'s you had to add, and give the logP.

of $-CH_3$'s required:_____

logP:_____

Appendix: What does logP mean?

Many researchers, especially drug designers, need to be able to estimate how hydrophobic a drug is. If it is too hydrophobic, it will not dissolve well enough in blood (which is mostly water) to get to the target. If it is too hydrophilic, it may have trouble passing through the hydrophobic core of the cell membranes. They could just make the drug and see, but synthesis is very expensive and they'd like to be able to at least estimate its hydrophobicity beforehand.

If they were able to make the drug, they would measure its hydrophobicity by adding it to a flask containing water and octanol ($H_3CCH_2CH_2CH_2CH_2CH_2CH_2CH_2OH$ – a very hydrophobic molecule). Since water and octanol don't mix appreciably, you get two layers. If the drug is very hydrophilic, you will find all of it in the water layer and none in the octanol. If the drug is very hydrophobic, you will find all of it in the octanol layer and none in the water layer. If the drug is in between, you will find some in the water and some in the octanol. The ratio of the amount found in the octanol divided by the amount found in the water is called the octanol-water partition coefficient; this is abbreviated P_{OW} and is higher the more hydrophobic a molecule is. Since P_{OW} varies over a large range, it is convenient to take the base-10 logarithm of P_{OW} or $\log(P_{OW})$.

For example, consider a drug that is moderately hydrophobic. Suppose that if you put 10 grams of the drug into the octanol/water flask, shake it up, and let it come to equilibrium, you find 9.09 grams of the drug in the octanol and 0.909 gram of the drug in the water. The P_{OW} = 9.09/0.909 or 10 (10 times more of the drug goes into the octanol than the water). The $\log(P_{OW})$ would be $\log(10)$ or 1. So, the logP would be 1, what you'd expect for a moderately hydrophobic molecule. The table below shows some other values.

logP	–2	–1	0	1	2
P_{OW}	0.01	0.1	1	10	100
% of molecule in octanol	0.99	9.09	50	90.9	99
% of molecule in water	99	90.9	50	9.09	0.99

The Molecular Calculator examines the structure you submit to it and estimates the $\log(P_{OW})$ using a variety of measured and calculated factors.

(2) MACROMOLECULES

(2.1) Lipids and phospholipids

(2.1.1) Organisms use fats and lipids as an energy reserve. Fats are important in transporting other nutrients such as the vitamins A, D, E, and K, which are not water soluble. Fats also form an essential part of the cell membrane. Some fatty acids, like those in Crisco or butter, form a solid at room temperature, whereas others, like those in corn oil, are liquid at room temperature. A saturated fatty acid contains no C=C bonds, as shown below.

$CH_3-(CH_2)_{14}-COOH$:

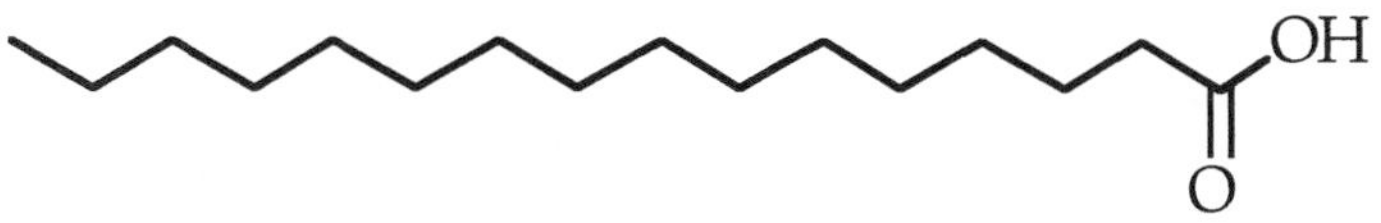

An unsaturated fatty acid has one or more C=C bonds.

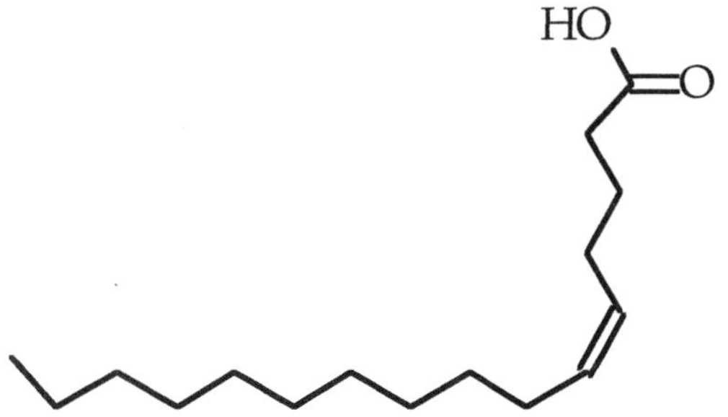

Which fatty acid do you predict will be solid at room temperature? Explain your answer.

(2.1.2) An example of a phospholipid is shown below. Phospholipids are a major component of _____.

A phospholipid contains both polar and nonpolar domains. Circle the polar domain. Box the nonpolar domain.

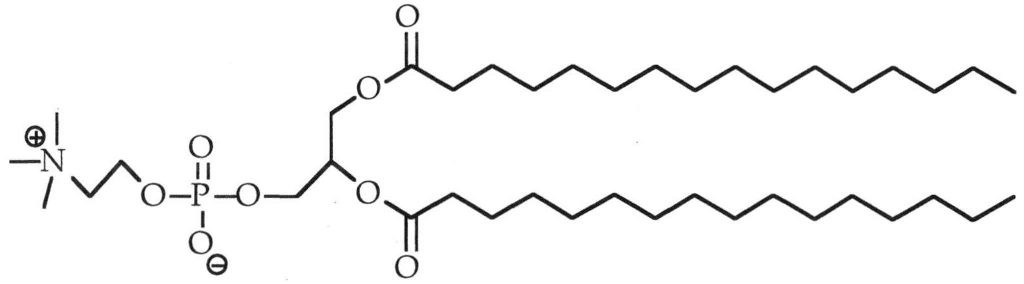

A schematic of a phospholipid can be drawn like this: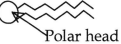
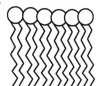
Polar head

Explain why you would not find phospholipids arranged like this ⟶
in the cell.

(2.1.3) Phospholipids can spontaneously form three different structures in aqueous environments. Draw the three possible structures that can be formed by phospholipids. Explain why the phospholipid molecules form these structures.

(2.2) Nucleic acids

A more in-depth treatment of nucleic acids can be found in Chapter 3.

(2.2.1) Consider the two molecules shown below. Which is DNA and which is RNA? Describe the purpose(s) each serves in the cell.

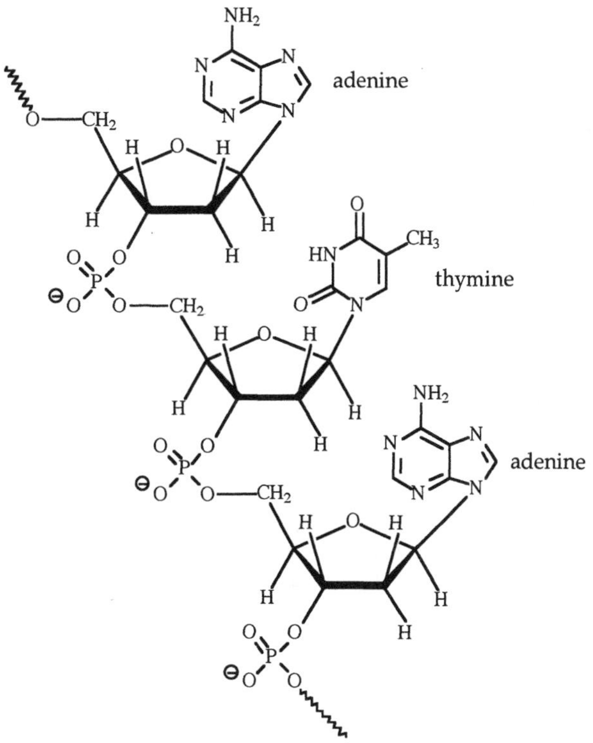

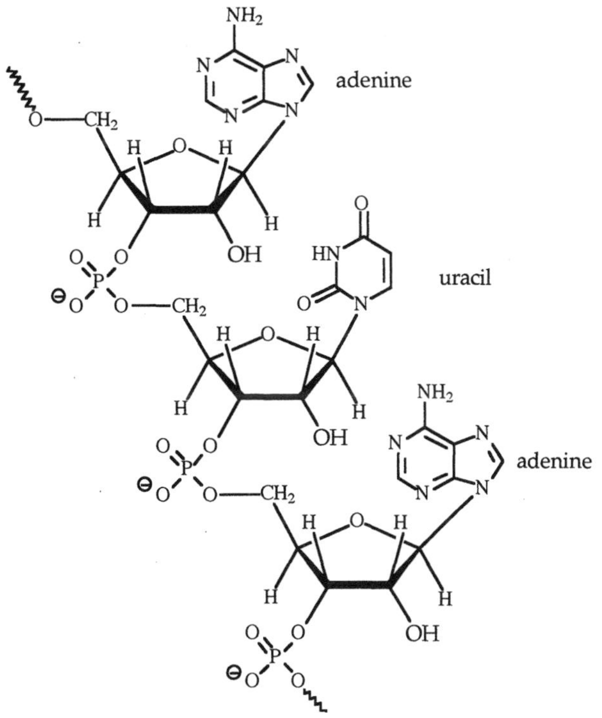

(2.3) Polypeptides and proteins, background

Diagnostic Problem:

Below is a small polypeptide.

a) It is composed of _____ amino acids.

b) Give the sequence of the amino acids in this polypeptide (the primary structure) and label the N and C termini.

c) Circle the peptide bonds. Are these bonds covalent or noncovalent?

d) For the pairs of amino acids given below, **circle** each side chain. Give the <u>strongest</u> type of interaction that occurs between the <u>side-chain</u> groups of each pair.

<table>
<tr><th>Amino Acids</th><th>Interaction</th></tr>
</table>

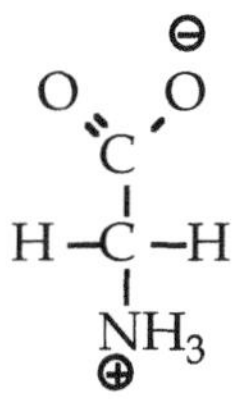

Glycine

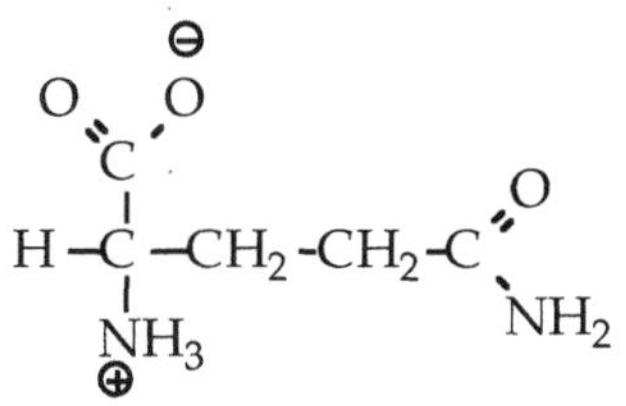

Glutamine

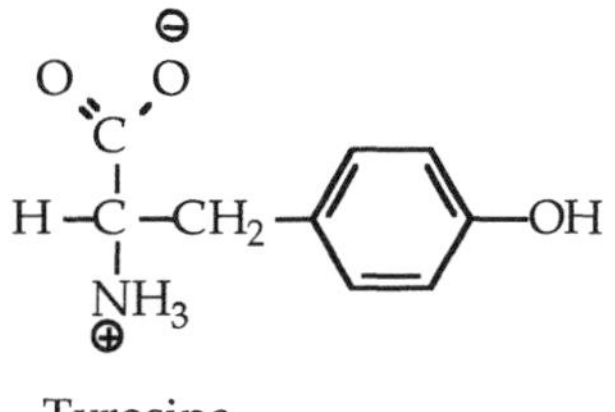

Tyrosine

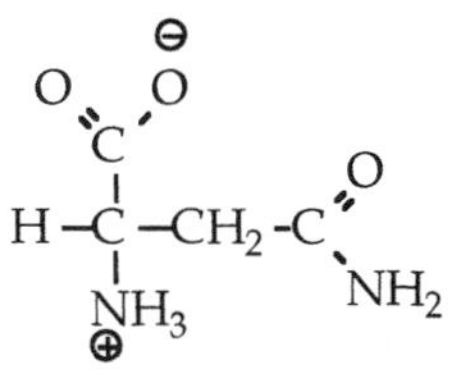

Asparagine

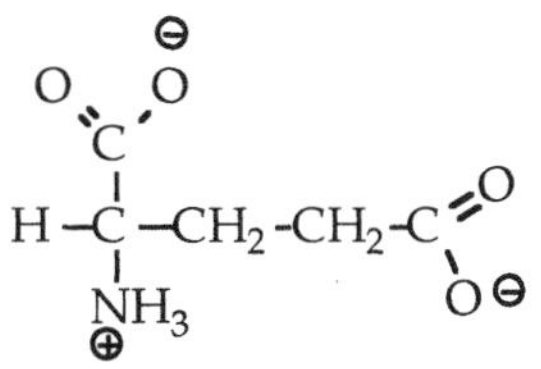

Glutamic Acid

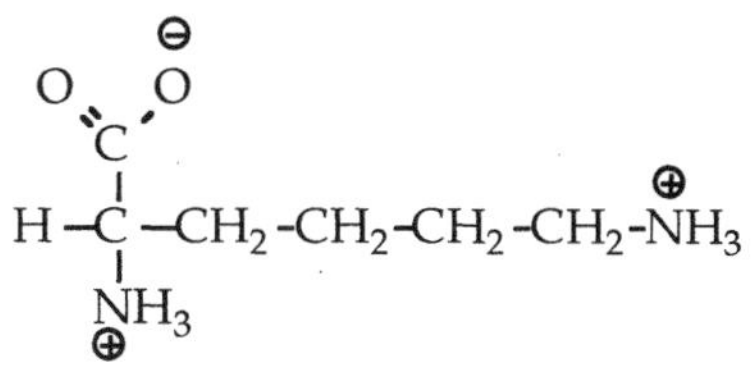

Lysine

e) What are the types of structural organization in a polypeptide?

Answer to Diagnostic Problem:

a) It is composed of *four* amino acids.

b) Give the sequence of the amino acids in this polypeptide and label the N and C termini.

   N-leucine-arginine-glutamic acid-asparagine-C

c) Circle the peptide bonds. Are these bonds (covalent) or noncovalent?

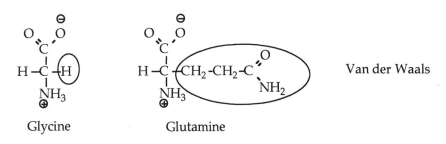

d) The side chains or R groups of the amino acids are circled, and the interactions described refer to interactions between the side-chain groups.

Amino Acids	Interaction

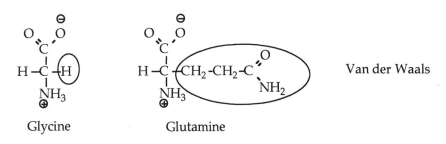

  Glycine   Glutamine   Van der Waals

Glycine is nonpolar, glutamine is polar, and the strongest interaction is van der Waals forces.

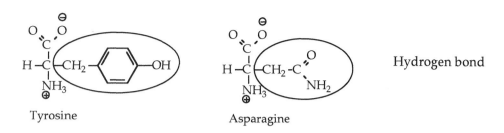

Tyrosine has a polar O–H group, asparagine is polar, and there is both a hydrogen donor and a lone pair of electrons, so a hydrogen bond could form.

Both side chains are polar and fully charged. One is positively charged, the other negatively charged, so an ionic bond could form.

e) What are the types of structural organization in a polypeptide?

- **Primary**: The linear order of the amino acids.
- **Secondary**: Regions of local structure (α-helix or β-sheet) mostly due to hydrogen bonding of one portion of the polypeptide backbone to another portion of the polypeptide backbone.
- **Tertiary**: The three-dimensional shape of a polypeptide.
- **Quaternary**: The interactions between subunits.

(2.3.1)

a) The structure of the amino acid glutamine is shown below.

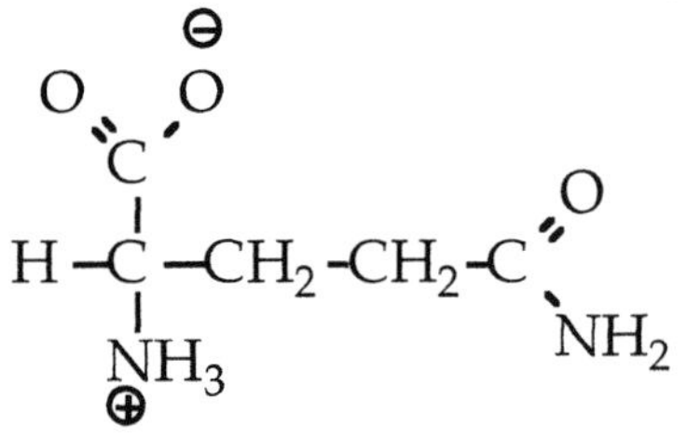

$$H-\underset{\underset{\oplus}{NH_3}}{\overset{\overset{\ominus}{\underset{\|}{C}}\overset{O\;\;O}{}}{C}}-CH_2-CH_2-C\overset{O}{\underset{NH_2}{}}$$

i) Give an amino acid whose side chain can form a <u>hydrogen bond</u> with the side chain of glutamine. There may be more than one correct answer here; give only one.

ii) Next to the structure of glutamine shown above, draw the side chain of the amino acid you selected in part (i) making a hydrogen bond with glutamine. Indicate the hydrogen bond with a dashed line. There may be more than one correct answer here; give only one.

b) The structure of the amino acid lysine is shown below.

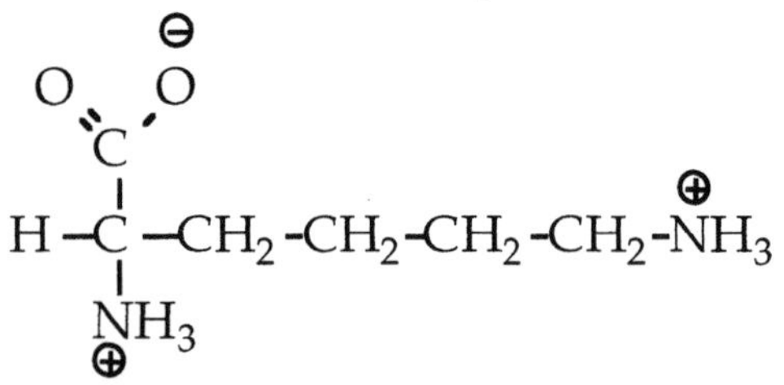

$$H-\underset{\underset{\oplus}{NH_3}}{\overset{\overset{\ominus}{\underset{\|}{C}}\overset{O\;\;O}{}}{C}}-CH_2-CH_2-CH_2-CH_2-\overset{\oplus}{NH_3}$$

i) Give an amino acid whose side chain can form an <u>ionic bond</u> with the side chain of lysine. There may be more than one correct answer here; give only one.

ii) Next to the structure of lysine shown above, draw the side chain of the amino acid you selected in part (i) making an ionic bond with lysine; indicate the ionic bond with a dashed line. There may be more than one correct answer here; give only one.

c) The structure of the amino acid leucine is shown below.

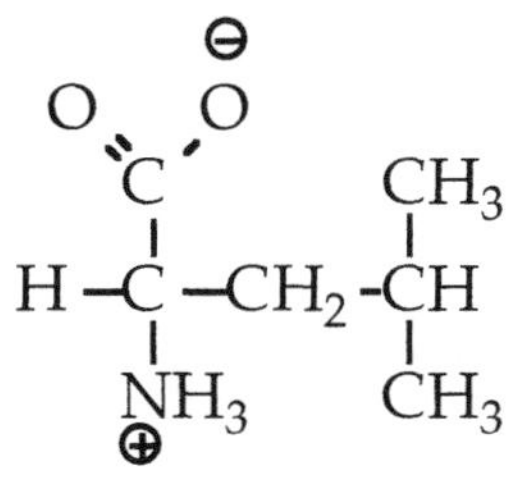

$$
\begin{array}{c}
O \overset{\ominus}{\underset{\diagdown}{}} O \\
\overset{\diagup}{C} \\
| \\
H-C-CH_2-\overset{CH_3}{\underset{CH_3}{\overset{|}{\underset{|}{CH}}}} \\
| \\
\underset{\oplus}{NH_3}
\end{array}
$$

   i) Give an amino acid whose side chain can form a <u>hydrophobic interaction</u> with the side chain of leucine. There may be more than one correct answer here; give only one.

   ii) Next to the structure of leucine shown above, draw the amino acid you selected in part (i) making a hydrophobic interaction with the side chain of leucine. Indicate the hydrophobic interaction by circling the interacting parts of the two side chains. There may be more than one correct answer here; give only one.

(2.3.2) Researchers have found that some bacteria communicate with one another by releasing small peptides into their growth media.

Consider the sequence of the peptide shown below:

   N-Val-Arg-Cys-Asn-C

Draw the structure of the peptide (including the side chains of each amino acid) as it would be found at pH 7.0.

(C5) Computer-Aided Problems 5
Because secondary structure is a 3-dimensional concept, there will be no problems on paper in this section.

Objectives:
- To observe the three major types of protein secondary structure.
- To see how they can be fitted together to form a protein.
- To introduce you to the complex 3-d structures of proteins.

Procedure:
1) Launch "Molecules in 3-d" in the "Biochemistry" folder and click on the tab for this problem "Lysozyme I." Then, click the "Load lysozyme and show backbone" button (note that it may take a few seconds to load the structure). You will see something like this:

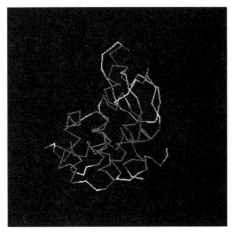

The image on the black screen shows the backbone of the lysozyme molecule. You can rotate or zoom in on this just as you did with the small molecules. Different amino acids have been colored based on their secondary structure.

- alpha helix = red
- beta sheet = yellow
- turn = blue (this is a specifically shaped turn of the backbone)
- random coil = white (none of the above)

In addition to being able to rotate and zoom in on a molecule, this program also allows you to identify the amino acid over which you have placed the cursor. This works much the same as it did for the earlier molecular visualization exercises. The program will then display information on the atom in a pop-up window (this does not always work on Macintoshes). The information in the pop-up window is rather cryptic.

This image shows "[TYR]161.CA #1273." This can be broken down into:
- "[TYR]" means that you clicked on a tyrosine.
- "161" means that the tyrosine you clicked on was amino acid number 161. Amino acids are numbered starting with #1, the amino terminus.
- "CA" means that you clicked on the alpha carbon of the lysine in the polypeptide chain.
- "#1273" means that the alpha carbon of amino acid #161 is atom number 1273, the overall protein molecule. This information is not particularly useful; do not confuse this number with the amino acid number.

You can also get this information by clicking on the amino acid of interest. Information about that amino acid will appear in the lower line of text under the structure window. For example, if you clicked on the part of the backbone shown above, you would see, "You just clicked: Amino acid TYR; Number 161" in the text line at the bottom of the "Molecules in 3-d" window.

Note that, when using either of these methods, it can be tricky to be sure what you have clicked on. Often, you can get a clearer "click" by rotating the molecule until the desired amino acid is clearly separated from the others.

a) Using this, describe the secondary structure of all the amino acids in the enzyme lysozyme.
- Start by finding one of the ends of the backbone chain. Interestingly, both ends are quite close together.
- Put the cursor over it or click on it. If it is number 1, you have found the amino terminus. Start here.
- Trace the backbone as it coils and twists. It may be difficult to be sure what you are clicking on; try rotating the molecule as you work. Determine the secondary structure of each amino acid.

Here is how it should look for the first 20:

#1 to #2: random coil   #3 to #11: alpha helix
#12 to #13: random coil   #14 to #20: beta sheet

You should complete this for the rest of the protein.

b) Click the button marked "Show alpha helix." You will see a short segment of alpha helix. You may need to zoom in to see it in detail (shift-drag up or down on the molecule). Each sharp bend in the backbone corresponds to one amino acid. Roughly how many amino acids are there per turn of the alpha helix? Hint: you may find it easier to count the number of amino acids in two or more turns.

c) Click the button marked "Show beta sheets." Beta sheets are composed of two or more parallel backbone segments. In some cases, the backbone segments run from amino to carboxyl terminus in the same direction ("parallel beta sheet"):

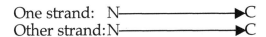

In other cases, the backbone segments run amino to carboxyl terminus in opposite directions ("antiparallel beta sheet"):

There are four regions where the backbone of lysozyme is in the beta sheet form where you can clearly see the interacting strands: 15 to 17, 24 to 27, 31 to 34, and 56 to 58. For each of the sections of beta sheet, determine which sections are interacting and whether they are parallel or antiparallel. You can find the direction of a given part of the protein by clicking on the amino acids; if the numbers increase, it means that you are moving toward the carboxyl terminus.

(2.4) Polypeptides and proteins, interactions

(2.4.1) Toxic Shock Syndrome Toxin (TSST) is a protein produced by the bacterium *Staphylococcus aureus*. During an *S. aureus* infection, the TSST protein binds to MHC Class II proteins (MHC II) found on the surface of antigen-presenting cells of the patient's immune system. Binding of TSST to MHC II results in hyperactivation of the immune cells, which leads to the symptoms of toxic shock syndrome. A simplified version of the structure of both proteins has been determined; part of the binding interface of TSST as it binds to MHC II is shown below. (Note that "gln$_{47}$" is shorthand for "the 47th amino acid, starting from the amino terminus, is a glutamine.")

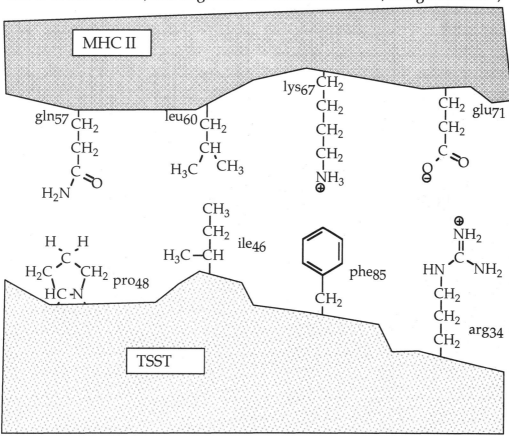

a) Classify each of the eight side chains shown above as "hydrophobic," "hydrophilic and charged," or "hydrophilic and polar."

b) Each side chain of MHC II interacts with an opposite side chain in TSST (for example, Gln$_{57}$ of MHC II interacts with Pro$_{48}$ of TSST). What type(s) of interactions (covalent, hydrogen, ionic, or van der Waals) are possible between side chains of the MHC II protein and the opposite TSST side chain?

MHC II side chains	Interaction with opposite side chain of TSST
Gln$_{57}$	
Leu$_{60}$	
Lys$_{67}$	
Glu$_{71}$	

c) Suppose you wanted to design an altered version of either MHC II or TSST that would make the interaction between TSST and MHC II stronger than in the normal situation. What amino acid would you change and what would you change it to? There are many possibilities; give one and explain how your change would strengthen the binding.

The remainder of this question deals with some hypothetical (possible but not yet studied) altered versions of the TSST protein and how they would interact with the MHC II protein.

d) Version 1 of TSST ($TSST_1$; normal TSST is called $TSST_{Norm}$) has a glutamine at position 34 instead of an arginine. Under conditions where $TSST_{Norm}$ would bind to MHC II, $TSST_1$ **does not bind**. Provide a reasonable explanation for why $TSST_1$ does not bind.

e) Version 2 of TSST ($TSST_2$) has a glutamic acid at position 34 instead of an arginine. Under conditions where $TSST_{Norm}$ would bind to MHC II, $TSST_2$ **does not bind**. Provide a reasonable explanation for why $TSST_2$ does not bind.

f) Version 3 of TSST ($TSST_3$) has a leucine at position 46 instead of an isoleucine. Under conditions where $TSST_{Norm}$ would bind to MHC II, $TSST_3$ **does bind**. Provide a reasonable explanation for why $TSST_3$ does bind.

g) Version 4 of TSST ($TSST_4$) has a glutamine at position 34 instead of an arginine **and** a serine at position 48 instead of proline. Under conditions where $TSST_{Norm}$ would bind to MHC II, $TSST_4$ **does bind**. Provide a reasonable explanation for why $TSST_4$ does bind.

h) Version 2 of TSST ($TSST_2$) does not bind to normal MHC II. What amino acid substitution could you make in MHC II that would allow it to bind to $TSST_2$? There are several possibilities; describe one and explain your reasoning briefly.

(2.4.2) Sickle-cell anemia is a genetic disease involving hemoglobin, the protein which carries O_2 in the red blood cells. The disease symptoms are caused by the presence of abnormal hemoglobin molecules (Hb^S, S for sickle-cell; normal hemoglobin molecules are designated Hb^+) which aggregate under certain conditions, preventing proper red blood cell function.

Aggregation of Hb^S begins with an interaction between two molecules of Hb^S; the resulting dimers then aggregate to form the disease-causing long polymers. The aggregation is driven by an interaction between the side chain of amino acid #6 (valine) of one hemoglobin molecule with a pocket formed by the side chains of amino acids #85 (phenylalanine) and #87 (leucine) of another hemoglobin molecule. This is shown below.

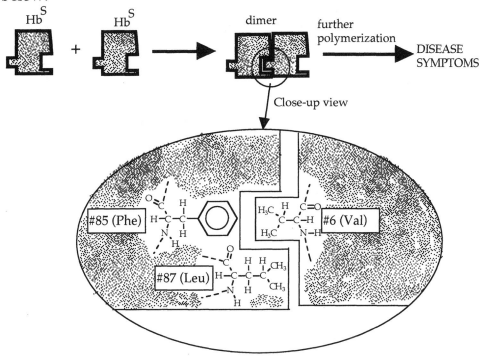

Only the three relevant amino acids are shown; the peptide backbone is indicated with a dashed bond.

a) What type of bond/interaction exists between the side chain of valine #6 and the side chains of phenylalanine #85 and leucine #87 (ionic bond, hydrogen bond, hydrophobic interaction)?

b) Wild-type hemoglobin does not form dimers or polymers of any kind. The only difference between wild-type (Hb$^+$) and sickle-cell (Hb^S) hemoglobins is:

- Amino acid #6 in Hb^S (sickle-cell) is valine (shown on the preceding page).

- Amino acid #6 in Hb$^+$ (wild-type) is glutamic acid.

Based on these data, provide a plausible explanation for why Hb$^+$ does not form polymers.

c) Consider the completely hypothetical case of a mutant form of hemoglobin that is identical to wild-type hemoglobin (Hb$^+$), except that amino acid #6 in the mutant hemoglobin (Hb^{Phe}) is phenylalanine instead of glutamic acid. There are two possibilities:

i) Suppose that Hb^{Phe} **does not** form polymers under any circumstances. Provide a plausible explanation for this observation, based on the structures of the molecules involved.

ii) Suppose that, under circumstances where Hb^S forms polymers, Hb^{Phe} **does** form polymers with the same general structure as polymers of Hb^S. Provide a plausible explanation for this observation, based on the structures of the molecules involved.

(2.4.3) The structure of the enzyme tryptophan synthetase has been studied extensively by a variety of methods. In a series of studies, Yanofsky and coworkers examined the effect on enzyme activity of various amino acid changes in the protein sequence (*Federation Proceedings*, **22**:75 [1963] and *Science* **146**:1593 [1964]). Altered amino acids are shown in **bold**. "Wild-type" is the normal strain isolated from the wild.

Strain	Amino Acid at Position A	Enzymatic Activity
wild-type	Gly	full
mutant 1	**Glu**	none
mutant 2	**Arg**	none

Here are two possible explanations for these results:

- The Gly ⇒ Glu and Gly ⇒ Arg changes introduce a **charge** (+) or (−) into a region of the protein that requires an uncharged amino acid like glycine.

- The Gly ⇒ Glu and Gly ⇒ Arg changes introduce much **larger** amino acid side chains into a space in the protein that requires a small amino acid like glycine.

Yanofsky and coworkers collected more mutants and examined their proteins to determine which of the above explanations was more likely to be correct:

Strain	Amino Acid at Position A	Enzymatic Activity
wild-type	Gly	full
mutant 3	**Ser**	full
mutant 4	**Ala**	full
mutant 5	**Val**	partial

a) Which of their models is supported by these data? Why?

Alterations of amino acids at another location in the protein were found to interact with alterations at position A.

Strain	Amino Acid at Position A	Amino Acid at Position B	Enzymatic Activity
wild-type	Gly	Tyr	full
mutant 1	**Glu**	Tyr	none
mutant 6	**Glu**	**Cys**	partial
mutant 7	Gly	**Cys**	none

b) Explain the behavior of mutant 6 in terms of your model of part (a).

c) Given your model above, explain the lack of activity found in mutant 7.

(2.4.4) Nucleosomes are protein complexes formed by eight interacting subunits. These complexes aid in the orderly packing of DNA by acting as a spool around which the DNA double helix is wound.

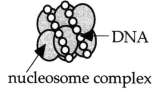

nucleosome complex

a) How many polypeptides compose the nucleosome complex?

b) What is quarternary structure? Does the nucleosome complex have quarternary structure?

c) The following sequence of amino acids is found as part of the primary structure of the nucleosome complex:

Val-Leu-Ile-Phe-Val-Val-Ile-Ile

i) In what general region of the nucleosome complex would you expect to find this stretch of amino acids?

ii) Why did you choose this region?

d) Some regions of the nucleosome complex have high percentages of lysine and arginine. Given the function of the nucleosome:

i) Where might these regions be found?

ii) What might be the role of these regions?

Your friend wants to examine the interactions between nucleosome complexes and DNA double helices. He prepares three identical samples of nucleosome complexes associated with DNA and treats each sample with an agent that disrupts a different type of molecular force. He disrupts hydrogen bonds in sample 1, ionic bonds in sample 2, and peptide bonds in sample 3.

You know that all of the treatments eliminate the binding between nucleosome complexes and DNA double helices and also disrupt other interactions.

e) Indicate how each treatment affects the nucleosome complexes by listing the appropriate number(s) in the table below. Choose <u>any</u> or <u>all</u> that apply.

    1) No change, complex intact.

    2) Disrupt tertiary structure.

    3) Disrupt disulfide bonds.

    4) Disrupt secondary structure.

    5) Disrupt primary structure.

Treatment	Effect on nucleosome complexes (list appropriate number(s) from above)
Disrupt hydrogen bonds	
Disrupt ionic bonds	
Disrupt peptide bonds	

f) Indicate how each treatment affects the structure of DNA double helices by listing the appropriate number(s) in the table below. Choose <u>any</u> or <u>all</u> that apply.

    1) No change, double helices intact.

    2) Disrupt base pairing.

    3) Disrupt hydrophobic interactions.

    4) Disrupt sugar-phosphate backbone.

Treatment	Effect on structure of DNA double helices (list appropriate number(s) from above)
Disrupt hydrogen bonds	
Disrupt ionic bonds	
Disrupt peptide bonds	

(2.4.5) Suppose you have two different protein α-helices that bind to one another. A variety of amino acids are seen at the binding interface between these helices. At the binding surface of helix 1 is a serine, an alanine, and a phenylalanine. On the binding surface of helix 2 is a glutamine, a methionine, and a tyrosine. (Note the binding interface in the figure below.)

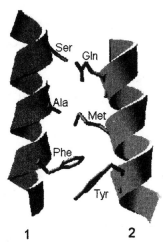

a) Interactions between these amino acids hold the helices together. What is the strongest possible interaction between <u>each</u> of the following pairs of amino acids? Choose from covalent bonds, van der Waals forces, ionic bonds, and hydrogen bonds.

	Amino acids	Strongest interaction
i)	Ser and Gln	
ii)	Ala and Met	
iii)	Phe and Tyr	

If a little heat is applied to these helical proteins, you observe that the helices no longer bind one another and instead are free helices in solution. If even more heat is applied, you no longer even see helices. Only elongated peptides with no defined structure are observed.

b) Explain why at low heat the proteins maintain a helical structure but fail to interact, while higher heat produces elongated peptides.

You replace both the phenylalanine of helix 1 and the tyrosine of helix 2 with cysteine.

c) How does changing both these residues to cysteine affect the stability of the interaction? Why?

(C6) Computer-Aided Problems 6

The problems in this section deal with the enzyme lysozyme. Lysozyme catalyzes the breakdown of bacterial cell walls. Lysozyme is used by the bacterial virus called bacteriophage T4 to break out of the host cell.

1) Hydrophobic/Hydrophilic

In general, you would expect to find amino acids with hydrophobic side chains in the interior of a protein and amino acids with hydrophilic side chains on the outside of the protein. In this problem, you will explore a simple real-world protein to see how these principles are applied in nature.

Start the "Molecules in 3-d" program in the "Biochemistry" folder. Click on the tab for this problem "Lysozyme II." Click the "Load lysozyme and show exterior; red = phobic" button; it may take a little while to load the structure. You should see a black window with a collection of red and white spheres displayed. The red and white spheres are individual atoms of the protein lysozyme. Atoms in amino acids with hydrophobic side chains are red; hydrophilics are white.

   a) Look at the view you just loaded. You should see the red and white protein. Use the mouse to rotate it to see all sides. How would you characterize the amino acid side chains on the surface (all hydrophobic, mostly hydrophobic, equally hydrophobic and hydrophilic, mostly hydrophilic, all hydrophilic)? How well does this fit with your expectations? Provide a plausible explanation for why this might be so.

   b) Click the button marked "Show interior; red = phobic" to show a brief animation. The display will rotate lysozyme to a specific position, pause briefly, and then show the interior of the enzyme. The view shows what you would see if you sliced the protein in a vertical plane parallel to the screen and removed the front section – like slicing an orange down the middle and looking inside. How would you characterize the amino acid side chains in the interior (all hydrophobic, mostly hydrophobic, equally hydrophobic and hydrophilic, mostly hydrophilic, all hydrophilic)? How well does this fit with your expectations? Provide a plausible explanation for why this might be so.

   c) Click the button marked "Show valines." The display will show most of the atoms in the protein as balls made of tiny yellow dots; this allows you to see through them into the interior of the protein. Several other atoms are shown as solid spheres; these are the atoms in the nine valines in the protein. The atoms in the valines are colored according to what element they are (see the color scheme on the web page).

Valine has one of the most hydrophobic side chains of any amino acid. Valine's side chain is composed entirely of carbon (gray) and hydrogen (not shown).

For each valine, determine (to the best of your ability) whether the side chain is inside the protein or exposed to the water at the protein's surface. The best way to do this is to pick an individual valine (you can identify which one it is by putting the cursor over it or by clicking on it as in previous problems), then rotate the protein carefully while trying to see if any of the side chain is not covered by yellow dots. If you can see parts of the side chain that are not covered by yellow dots, then that side chain is exposed to the water surrounding the protein. If there is no way to see the side chain without looking through yellow dots, then the side chain is buried.

How many of the nine valines are completely buried? How does this match with your expectations? Why might this be so?

d) Click the button marked "Show lysines." The display will show the bulk of the atoms in the protein as balls made of tiny yellow dots; this allows you to see through them into the interior of the protein. Several other atoms are shown as solid spheres; these are the atoms in the 13 lysines in the protein. The atoms in the lysines are colored according to what element they are (see the color scheme on the web page).

Lysine has one of the most hydrophilic side chains of any amino acid. Lysine's side chain ends with a single positively charged nitrogen atom (blue).

For each lysine, determine (to the best of your ability) whether the side chain is inside the protein or exposed to the water at the protein's surface. You should focus on the most hydrophilic part – the blue nitrogen atom at the tip of the side chain. You can use the same method you used for part (c).

How many of the 13 lysines are completely buried? How does this match with your expectations? Why might this be so?

2) Side-chain interactions

We will next consider interactions between side chains of different amino acids in the protein lysozyme. These interactions contribute to the tertiary structure of the protein.

Start the "Molecules in 3-d" program in the "Biochemistry" folder. Click on the tab for this problem "Lysozyme III." Click the "Load Lysozyme" button. You must click this button first to load the structure for the other parts of this problem.

You will see a black window with the protein lysozyme shown in "ball and stick" mode – atoms are shown as balls and the covalent bonds connecting them are shown as rods. You can click on the "Show atoms as spacefill" button to change the representation "Spacefill" where atoms are shown as solid spheres at their actual sizes. You will find it useful to switch back and forth between the two views. Note that you may sometimes need to click this button three times to get the view to change.
- The ball and stick view shows covalent bonds as rods and is most useful for determining which atoms are *covalently bonded* to each other. The small size of the atoms can sometimes make it hard to tell which atoms are close together for noncovalent interactions.
- The spacefill view shows atoms as joined spheres of their approximate actual size in the molecule. It is most useful for determining which atoms are *closest together*. Because it does not show covalent bonds, it can sometimes be hard to figure out what atoms are covalently or noncovalently bonded.

There are several important things to note about these views:
- They show only the covalent bonds; you must *infer* the noncovalent bonds based on your knowledge of amino acids and their properties.
- They do not show hydrogen atoms; you must *infer* where they are based on your knowledge of amino acid structure. This is especially important when exploring hydrogen bonds.
- These views show only the amino acids listed on the button; the remaining amino acids are shown as dark lines.

Because these problems deal with amino acids in an actual protein, it is important to consider the relative positions and conformations of the side chains. Put another way, "Even if a particular interaction is possible based on the structures on paper alone, the side chains must be arranged properly in order for the interaction to actually occur in the protein."

Each part of this problem involves looking at the interaction between the side chains of two amino acids.

For each problem, click the appropriate button and answer the following questions. You will find it useful to rotate, zoom in, and/or change from ball and stick to spacefill views. The questions are:

i) Look up the structures of each amino acid in your textbook. Based on these structures only, what interaction(s) are possible between their side chains?
- Ionic bond
- Hydrogen bond
- Hydrophobic interaction
- van der Waals interaction

ii) Which of the interactions you selected above is the strongest?

iii) Look at how these side chains are arranged in lysozyme and sketch their relative arrangement on paper. Be sure to add in the hydrogen atoms.

iv) Based on the structure you drew in (iii), what is the strongest interaction between the side chains in the actual protein?

a) Glu_{11} and Arg_{145} – click the button labeled "Show Glu 11 and Arg 145" and answer the four questions above.

b) Asp_{10} and Tyr_{161} – click the button labeled "Show Asp10 and Tyr 161" and answer the four questions above.

c) Gln_{105} and Trp_{138} – click the button labeled "Show Gln 105 and Trp 138" and answer the four questions above.

d) Met_{102} and Phe_{114} – click the button labeled "Show Met 102 and Phe 114" and answer the four questions on page 134.

e) Tyr_{24} and Lys_{35} – click the button labeled "Show Tyr 24 and Lys 35" and answer the four questions on page 134. This is a challenging one.

3) Effects of mutations on protein structure

In a truly heroic series of experiments, Rennell, Bouvier, Hardy, and Poteete (*Journal of Molecular Biology* **222**:67-87 [1991]) generated a huge set of mutant versions of lysozyme. In each individual mutant, only one amino acid was changed; all the others were the same. Each individual mutant was checked to determine whether it had full activity. In their studies, each of the 164 amino acids in lysozyme was individually changed to 13 alternatives.

We have chosen mutants that affect the amino acids you explored in problem (C2). In addition, the "Molecules in 3-d" program in the "Biochemistry" folder on the CD-ROM also contains a set of views of lysozyme specifically arranged for this problem "Lysozyme III."

Provide a plausible explanation for each of the following results in terms of your findings from problem (C2), keeping in mind the properties of different amino acid side chains. This first is given as an example:

> Question: "If Glu_{11} is changed to Ser, the resulting protein is <u>not</u> fully active."
> Complete answer: "Based on problem (C2), Glu_{11} normally makes an ionic bond with Arg_{145}. If the Glu at position 11 were replaced with Ser, an ionic bond would no longer be possible. Although an H-bond is possible, this would be weaker than an ionic bond. This weaker bond must not be strong enough to hold the protein in the correct shape; thus it is nonfunctional."

Your answers should be structured similarly.

a) If Glu_{11} is replaced with Arg, the resulting protein is <u>not</u> fully active.

b) If Glu_{11} is replaced with Phe, the resulting protein is <u>not</u> fully active.

c) If Glu_{11} is replaced with Asp, the resulting protein is <u>not</u> fully active.

d) If Arg_{145} is replaced with Ser, the resulting protein is <u>not</u> fully active.

e) If Arg_{145} is replaced by His, the resulting protein is <u>fully active</u>; if it is replaced by Lys, the resulting protein is <u>not</u> fully active.

f) If Tyr_{161} is replaced with Ser, the resulting protein is <u>not</u> fully active.

g) If Asp_{10} is replaced with Glu, the resulting protein is <u>fully active</u>.

h) If Gln_{105} is replaced with Glu, the resulting protein is <u>fully active</u>.

i) If Gln_{105} is replaced with Leu, the resulting protein is <u>fully active</u>. Why is this surprising?

j) If Met_{102} is replaced with Glu, Arg, or Lys, the resulting protein is <u>not</u> fully active.

k) Lys_{35} can be replaced with any amino acid and <u>all</u> the resulting proteins are <u>fully active</u>.

l) If Phe_{67} is replaced with Pro, the resulting protein is <u>not</u> fully active. You should go back to "Molecules in 3-d" problem "Lysozyme III" and look at the view of Phe_{67} and the secondary structure of the protein (click the button "Show Phe 67 and secondary struct."). In this view, Phe_{67} is shown as spheres; the rest of the protein is shown as backbone only. The backbone of Pro is slightly but significantly different from the backbone of all the other amino acids; you should check your textbook for details.

(2.5) Polypeptides and proteins, binding sites

One of the most important functions of proteins is to act on smaller molecules. Proteins do this by binding these smaller molecules via the noncovalent interactions we have already discussed.

(2.5.1) You are studying a protein, Protein A, which binds a small molecule, Molecule X. The binding site is shown below with Molecule X bound.

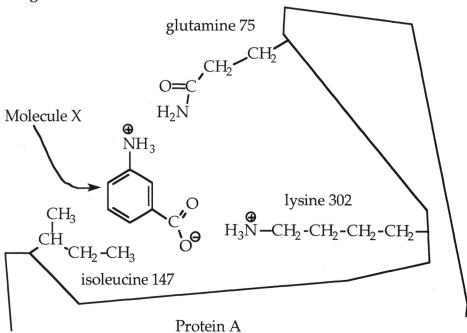

Molecule X binds to Protein A via the side chains of three amino acids in Protein A: glutamine 75, isoleucine 147, and lysine 302.

a) What is the <u>strongest</u> possible interaction between the side chain of **glutamine 75** and the nearest part of Molecule X?

b) What is the <u>strongest</u> possible interaction between the side chain of **isoleucine 147** and the nearest part of Molecule X?

c) What is the <u>strongest</u> possible interaction between the side chain of **lysine 302** and the nearest part of Molecule X?

Consider each of the following changes to the protein <u>separately</u>.

d) A mutant version of Protein A differs from the normal Protein A in only one amino acid: glutamine 75 is replaced by asparagine. This mutant protein no longer binds Molecule X. Explain why this change has this effect.

e) A different mutant version of Protein A differs from the normal Protein A in only one amino acid: lysine 302 is replaced by glutamic acid. This mutant protein no longer binds Molecule X. Explain why this change has this effect.

(2.5.2) Shown on the right is a hypothetical substrate molecule binding to a hypothetical protein. The substrate binds to the enzyme via noncovalent (hydrogen, ionic, hydrophobic) interactions. Below is a close-up of the substrate and substrate-binding region of the protein.

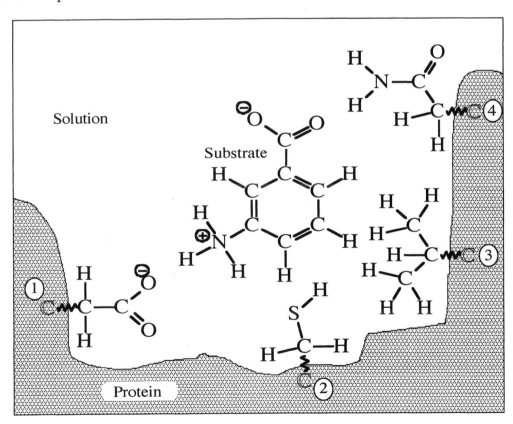

a) Each of the numbered groups is a side chain of a particular amino acid. For each side chain, give the amino acid.

b) Each of the four side chains of the protein that interact with the substrate are numbered on the figure above. For each side chain, state which type(s) of interactions it could have with the substrate in the configuration shown above. Also classify each side chain as hydrophobic, hydrophilic-polar, or hydrophilic-charged.

c) One way to study the noncovalent interactions between substrate and protein is to synthesize molecules similar to the substrate and see if they bind to the protein. Shown below are three "substrate analogs" along with the normal substrate. Explain, in terms of the interactions you described in part (a), why each of the analogs binds or fails to bind to the protein.

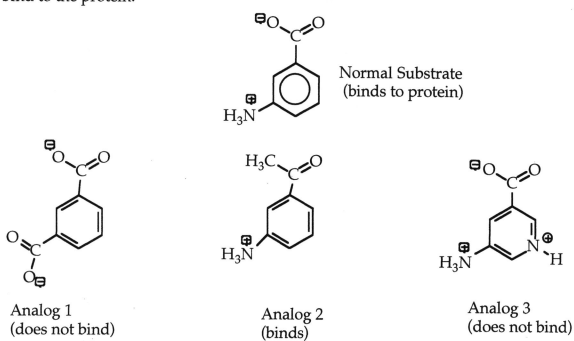

Normal Substrate
(binds to protein)

Analog 1
(does not bind)

Analog 2
(binds)

Analog 3
(does not bind)

d) Suppose you wanted to strengthen the binding of the substrate to the original enzyme by altering one of the four amino acids that you labeled in part (a) above. Which amino acid would you change, what would you change it to, and why?

(C7) Computer-Aided Problems 7

It is very important for cells to be able to repair DNA once it gets damaged. There are many mechanisms that cause DNA damage; one of these is "alkylation," where a foreign molecule becomes covalently attached to the DNA. Cells have specific enzymes that recognize this damaged DNA and begin the process of repair; one of these is alkyl adenine glycosylase (AAG). AAG recognizes adenine (A) bases in DNA that have inappropriate molecules covalently bonded to them. This enzyme and its interaction with DNA will be used as an example of enzyme-substrate interactions in 3-d.

a) Start the "Molecules in 3-d" program in the "Biochemistry" folder; it may take a little while to start up. Click on the tab for this problem "Enzyme AAG." Click the "Load AAG and DNA" button; it may take a little while to load the structure. You will see a view of AAG binding to some damaged DNA. You should rotate it to see all the parts of the molecule:

- The two DNA strands (the intertwined purple and yellow strands)
- The protein (light blue or gray)
- The damaged DNA base (red – buried at the interface between the DNA and protein)

Notice how the protein grabs onto the DNA and how the damaged DNA base is recognized by a pocket on the surface of the protein. The remainder of the problem will focus on the interactions between the DNA and the protein.

The next four parts involve interactions between particular side chains of the protein and particular parts of the DNA. The amino acids in the protein are identified as they have been throughout this book. The DNA bases are designated similarly: "T8" means "the 8th base in a particular chain, which is a Thymine (T)." Consult your textbook for structures of DNA and related molecules. When you click a particular button, the selected parts of the protein and DNA are shown as spheres connected by lines; the rest of the protein and DNA are shown as dark gray lines.

The program includes one important feature to make it easier for you to solve these problems: "Show atoms as spacefill" and "Show atoms as ball and stick" buttons – these switch back and forth between "spacefill" and "ball and stick" representations. For descriptions of "spacefill" and "ball-and-stick," see problem (C6) part (2).

For each of the interacting parts below, you should do the following:

 i) Click the corresponding button (for example, for part (a), click "Show Arg 182 in the protein and T 8 in the DNA"). Draw the interacting parts of the amino acid and the DNA. You do not have to draw the complete amino acids or nucleotides; draw only the few atoms that are interacting and their immediate neighbors.

 ii) What are the possible interactions between these parts of the molecule? Which is the strongest?

b) Arg_{182} and T_8

c) Thr_{143} and G_{23}

d) Met_{164} and T_{19}

e) Tyr_{162} and T_8

The next parts involve particular mutations in AAG. These were explored in another large study by Lau, Wyatt, Glassner, Samson, and Ellenberger (*Proceedings of the National Academy of Sciences of the United States* **97**(25):13573-13578 [2000]). In these experiments, they determined which amino acid substitutions are "tolerated." That is, the set of mutations that result in AAG proteins that are still fully active.

For each mutation, use your findings from previous parts (b) through (e) and your understanding of amino acids and protein structure to provide a plausible explanation for these observations. Here is an example:

 Statement: "No substitutions of Arg_{182} are tolerated."

 <u>Complete explanation</u>: "Arg has a medium-length side chain with a (+) charge as well as H-bond donors and acceptors; it cannot make a hydrophobic interaction. It makes an ionic bond with the (–)-charged oxygen on the phosphate group of T_8 in the DNA. Both Lys and His should be capable of making an ionic bond with a negative oxygen atom. Since neither His nor Lys is tolerated at this position, the amino acid at this position must be a specific size as well as (+)-charged."

f) Thr$_{143}$ can be replaced by Gln to produce a fully functional protein.

g) Met$_{164}$ can be replaced by Ile or Phe to produce a fully functional protein.

h) No substitutions of Tyr$_{162}$ are tolerated.

(C8) Computer-Aided Problems 8
This problem is designed to challenge you; it will show you how proteins interact with small molecules, how changes in structure influence binding, and how drugs can be designed based on structural information.

Aspirin and related drugs act to prevent pain by inhibiting the enzyme cyclo-oxygenase 2 (COX-2). Aspirin was discovered based on folk remedies for pain and inflammation. The latest generation of these drugs has been custom designed based on the structures of the molecules involved. More details can be found in problem 3.1.7. This problem deals only with the binding of the drug Celebrex and related drugs to the enzyme COX-2.

a) Start the "Molecules in 3-d" program in the "Biochemistry" folder. Click on the tab for this problem "Enzyme COX-2." Click the "Load COX-2 with drug bound" button; it may take a little while to load the structure. You will see a view of COX-2 with the drug bound.

This is the largest and most complex enzyme we have looked at so far. COX-2 is active as a dimer – two identical protein chains associated via noncovalent bonds. This is an example of a protein with quaternary structure.

The drug is embedded deeply in the protein. If you show the protein as solid spheres, only a small part of the drug is visible from the surface. If you click the "Show protein atoms as dots" button, you can see how buried the drug is. Notice also that the drug is much larger than the hole it had to pass through to get into the protein. This illustrates an important part of protein structures: to varying degrees, all proteins "breathe," partly unfolding and then refolding all the time. Presumably, the drug molecule got inside the protein during one of the partly unfolded states.

The next parts of the problem deal with altered versions of the drug and their binding properties. Because COX-2 is such a large and complex enzyme and the drug is so deeply buried, the views in this section are greatly simplified. They show only the drug and the side chains of the amino acids that contact the drug directly. You can change the drug and/or the protein to dots to help you see the important parts of the molecules. You will find that you need to switch back and forth often to get a clear picture. You will also need to click on atoms in the protein to identify the amino acids involved. To see these views, click the tab for "COX-2 Inhibitors" and click the "Load COX-2 with model drug bound" button.

Another note: this problem involves molecules that contain fluorine (F), chlorine (Cl), bromine (Br), and iodine (I). In the context of this problem, these atoms make only one covalent bond and cannot form hydrogen bonds. Bonds between these atoms and carbon are slightly more polar than carbon-carbon bonds, but not as polar as C–N or C–O bonds. For our purposes, it is sufficient to consider these atoms to be medium sized (in order from smallest to largest: -H, -F, -Cl, -Br, -I, -CH₃) and hydrophobic.

b) The structure of the model drug is shown below with the approximate positions of the side chains of COX-2 indicated:

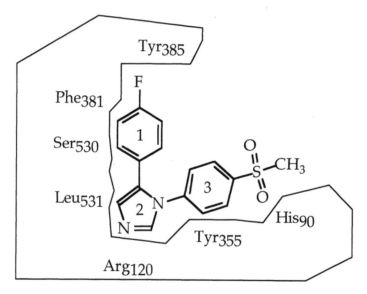

Based on the structures of the amino acids shown, give the strongest possible bond between the closest parts of the side chain and the model drug.

These next problems deal with derivatives of the model drug shown previously. These are chemical relatives of the drug that differ in small ways. They are described in a paper by Almaransa et al. (*Journal of Medicinal Chemistry* **46**:3463-3475 [2003]). These derivatives have changes at four parts of the model drug. These four places, marked A through D, are shown below:

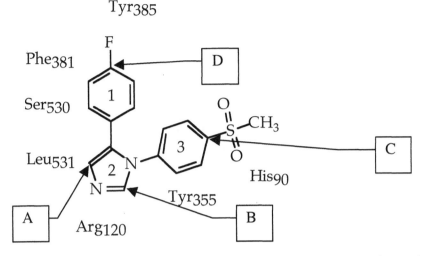

You can highlight parts A, B, C, and D by clicking the buttons to the right of the display of the molecule.

c) The first set of molecules is based on the structure shown above. That is, for Molecule 1 (shown above):

A = -H
B = -H
C = -SO$_2$CH$_3$
D = -F

They then tested molecules with different atoms covalently bonded to the carbon marked A to see if they would bind to COX-2. The results are as follows:

Molecule	Group Attached to Carbon A	Binds to COX-2?
Molecule 1	-H	no
Molecule 2	-Cl	yes
Molecule 3	-Br	yes
Molecule 4	-CH$_3$	yes
Molecule 5	-CH$_2$CH$_3$	no

In later parts of this problem, we will ask you to explain results like these. Here, we give our analysis as an example.

    i) <u>Question</u>: "Explain these results in terms of the structures of the drug molecules and COX-2."

<u>Complete answer</u>: All of the groups are hydrophobic; they differ only in their size. Therefore, size must be the relevant factor at position A. These results show that the group attached to carbon A must be medium-sized, neither too small nor too large. Only those molecules (2 through 4) have groups at position A that are neither too small (1) nor too large (5).

ii) Based on the answer we gave for part (i), would you expect the following version of the drug to bind to COX-2? Explain your reasoning.

$A = -CH_2CH=CH_2$
$B = -H$
$C = -SO_2CH_3$
$D = -F$

d) The researchers then based their next set of molecules on Molecule 2 because it bound well to COX-2. As a reminder, Molecule 2 has these groups:

$A = -Cl$
$B = -H$
$C = -SO_2CH_3$
$D = -F$

They then varied group B and got these results:

Molecule	Group Attached to Carbon B	Binds to COX-2
Molecule 2	-H	yes
Molecule 6	-Cl	no
Molecule 7	$-CH_2OH$	no

i) Explain these results in terms of the structures of the drug molecules and COX-2.

ii) Based on your answer to part (i), would you expect the following version of the drug to bind to COX-2? Explain your reasoning.

$A = -Cl$
$B = -CH_3$
$C = -SO_2CH_3$
$D = -F$

e) Since none of the derivatives in part (d) worked better than the one from part (c), the researchers started over from Molecule 2:

$A = -Cl$
$B = -H$
$C = -SO_2CH_3$
$D = -F$

They then tried changing the group attached to the carbon at location C to $-SO_2NH_2$. This version (Molecule 8) also bound to COX-2.

    i) Explain this result in terms of the structures of the drug and COX-2.

    ii) Would you expect the following version of the drug to bind to COX-2? Explain your reasoning.

$A = -Cl$
$B = -H$
$C = -H$
$D = -F$

f) The researchers then tried more versions based on the most successful version of the drug so far (Molecule 2):

$A = -Cl$
$B = -H$
$C = -SO_2CH_3$
$D = -F$

They varied the atoms attached to carbon D and got the following results:

Molecule	Group Attached to Carbon D	Binds to COX-2?
Molecule 2	$-F$	yes
Molecule 9	$-H$	yes
Molecule 10	$-CH_3$	yes
Molecule 11	$-OCH_3$	yes
Molecule 12	$-SCH_2CH_3$	yes
Molecule 13	$-OCH_2CH_2CH_3$	no

    i) Explain these results in terms of the structures of the drug molecules and COX-2.

ii) Based on your answer to part (i), would you expect the following version of the drug to bind to COX-2? Explain your reasoning.

A = -Cl
B = -H
C = -SO$_2$CH$_3$
D = -OCH$_2$CH$_3$

g) Click on the button "Load COX-2 with Celebrex bound" to show the drug Celebrex bound to COX-2. This drug was developed using a process like the one in the problem above and is now being successfully used to treat severe arthritis pain.

i) The structure of Celebrex is shown below:

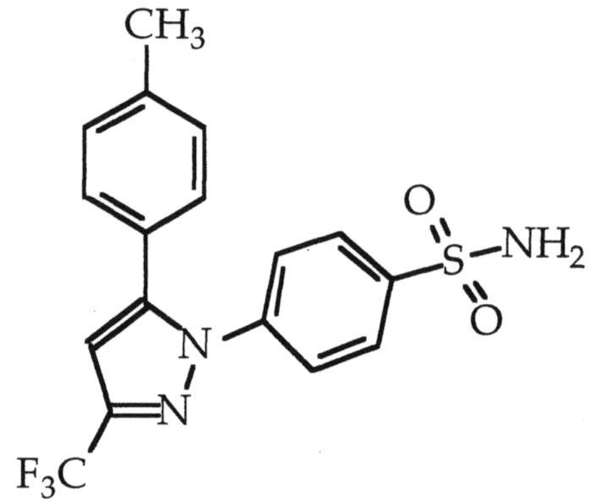

To which version of the model drug does Celebrex most closely correspond? Specifically, which groups at A, B, C, and D?

ii) Based on the data above, why would you predict that Celebrex would **not** bind to COX-2? Propose a plausible reason for why it **does** bind in real life.

(3) ENERGY, ENZYMES, AND PATHWAYS

(3.1) Energy and enzymes

Diagnostic Problem:

Graphs like that below are called "reaction coordinate diagrams"; they are one way of following the energetics of a reaction from start to finish. Energy of a compound at any given point is "G," and the free energy "ΔG" of a reaction or portion thereof is the difference between two points ($\Delta G = G_{product} - G_{reactant}$). For more on this, see your textbook.

Under standard conditions, the reaction $A + B \Rightarrow C + D$ has a positive ΔG_0.
The reaction $C + D \Rightarrow A + B$ has a negative ΔG_0.

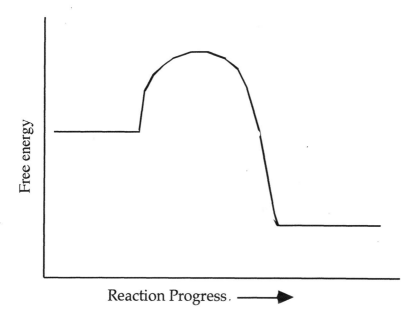

a) On the energy diagram above, label the following:
  • Activation energy
  • ΔG_0
  • C + D
  • A + B

b) Modify the diagram above to show how it would change if an enzyme that catalyzed the reaction is added.

c) What effect, if any, would adding an enzyme that catalyzed this reaction have?

Solutions to Diagnostic Problem:

a) and b)

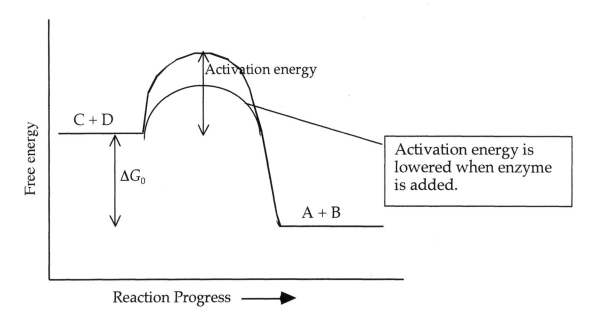

c) What effect, if any, would this have on the reaction?
An enzyme does not change the ΔG_0 of the reaction, but by lowering the activation energy it speeds up the reaction $C + D \Rightarrow A + B$.

(3.1.1) Virtually all of the reactions we will consider in this book will have multiple steps and several intermediates.

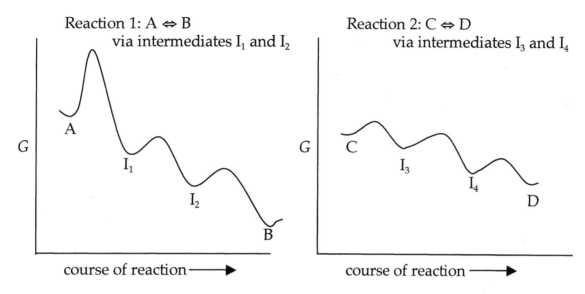

Reaction 1: A ⇔ B
via intermediates I_1 and I_2

Reaction 2: C ⇔ D
via intermediates I_3 and I_4

a) Which of the above reactions (A ⇒ B or C ⇒ D) yields the most free energy? Which reaction is the most thermodynamically spontaneous? Explain your reasoning.

b) We have described free energy as a way of predicting the likelihood of a reaction. However, even if thermodynamics says that a reaction is spontaneous (ΔG is negative), it is not necessarily true that the reaction will be fast (low activation energy). Given the diagram above, which reaction (A ⇒B or C ⇒D) will proceed at the greater rate? Explain your reasoning.

c) Would adding an enzyme that catalyzes the first step of reaction 1 (A ⇒ B) change the spontaneity (ΔG) and rate (activation energy) of the reaction A ⇒ B? Explain your reasoning.

(3.1.2) The first step in the pathway of glycolysis is the transfer of a phosphate group from ATP to glucose, giving glucose 6-phosphate (reactions of this type are phosphorylations). This step is shown in Reaction 1.

Reaction 1: Glucose + P_i $\Rightarrow$ Glucose 6-phosphate $\Delta G_0 = +3.3\,kcal/mol$

a) Under standard conditions, which is more favorable energetically, breakdown of glucose 6-phosphate (net reaction to the left) or formation of glucose 6-phosphate (net reaction to the right)? Why?

Reaction 2 is the hydrolysis of ATP to ADP:

Reaction 2:   ATP $\Rightarrow$ ADP + P_i $\Delta G_0 = -7.3\,kcal/mol$

b) Under standard conditions, which is more favorable energetically, formation of ATP (net reaction to the left) or breakdown of ATP (net reaction to the right)? Why?

c) Using the two reactions given, draw the overall reaction that makes the formation of glucose 6-phosphate a favorable, spontaneous reaction. In other words, what reaction do you add to (1) to make a reaction where glucose 6-phosphate is one of the products and the ΔG is (–)? What is the ΔG_0 for the overall reaction? In which direction does the reaction proceed?

(3.1.3) Given that the reaction $H_2 + O_2 \Rightarrow H_2O$ is <u>spontaneous</u>, answer the following questions.

a) Rank the following states of hydrogen and oxygen in order from the <u>highest free energy</u> to the <u>lowest free energy</u>.

State	Description
1	H_2O molecules
2	Free H and O atoms not bonded to anything
3	H_2 and O_2 molecules

Highest free energy_____

Middle free energy_____

Lowest free energy_____

b) What type(s) of bonds are being broken when you go from state (3) to state (2)?

covalent bonds   ionic bonds   hydrogen bonds   hydrophobic interactions

c) The reaction $H_2 + O_2 \Rightarrow H_2O$ proceeds only very slowly at room temperature. Draw a reaction coordinate diagram for this reaction on the graph below. Be sure to include:

- $H_2 + O_2$
- H_2O
- transition state
- appropriate line(s) connecting the three states

Note that only the <u>relative</u> levels are important here; you do not need to worry about the spacing between the levels.

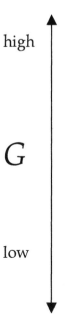

high

G

(3.1.4) Fresh raw potatoes contain an enzyme that rapidly produces an unwanted discoloration when the potato is peeled. This reaction can be thought of as:

$$O_2 + \text{aromatic amino acids (colorless)} \Rightarrow \text{cross-linked aromatic molecules (brown)}$$

Provide a plausible explanation for each of the following phenomena.

a) Putting cut potatoes under water slows the rate of browning. Previously submerged potatoes brown at the usual rate once they have been removed from the water.

b) Cooked (heated above 100°C for several minutes) potatoes do not brown at all even when they cool down (like in potato salad).

c) Sprinkling lemon juice over potatoes prevents browning, but sprinkling plain water does not. Hint: lemon juice has a very low pH.

(3.1.5) Transpeptidation is an essential step in the synthesis of the cell walls of certain bacteria. It is catalyzed by the enzyme transpeptidase. The active site of transpeptidase can also bind the β-lactam ring of penicillin (shown below). When this happens, the β-lactam ring opens and covalently binds to transpeptidase, permanently inactivating the enzyme.

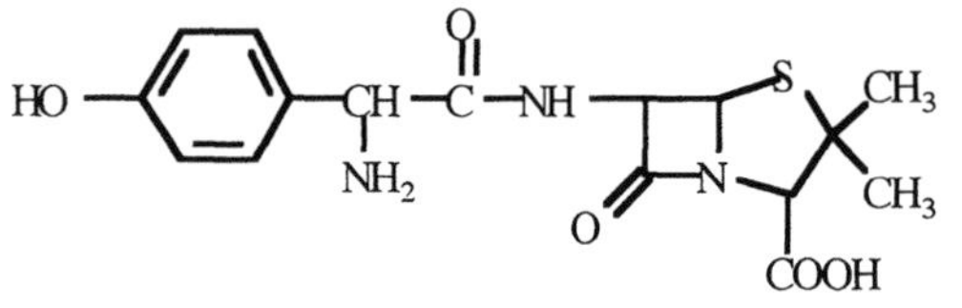

Penicillin G 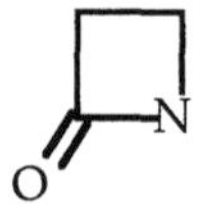   β-lactam ring

a) Based on this information and the structure of bacteria (consult your textbook), (i) explain the role of transpeptidase in bacterial cells, and (ii) how penicillin results in bacterial killing.

b) Based on this information and the structure of human cells (consult your textbook), why doesn't penicillin kill human cells?

c) Certain strains of bacteria contain an enzyme, β-lactamase, that catalytically opens the β-lactam ring of penicillin, rendering the penicillin nontoxic to bacteria. β-Lactamase therefore renders these cells resistant to penicillin. Transpeptidase is located outside the cell membrane, in the cell wall. Where do you expect β-lactamase to be located? Explain briefly, based on the function of β-lactamase in the cell.

(3.1.6) Insects have an enzyme that catalyzes the following reaction:

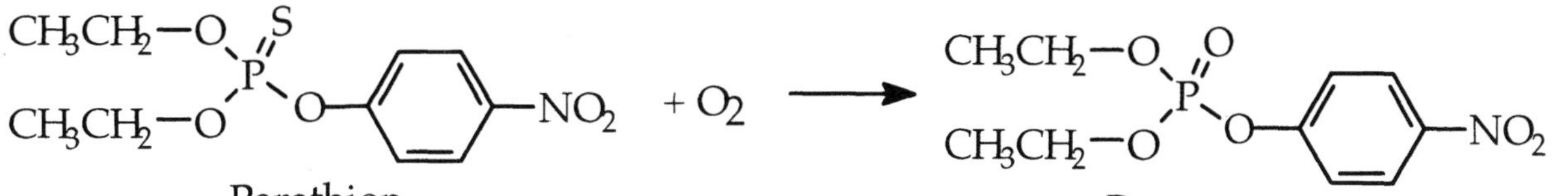

Parathion + O_2 $\longrightarrow$ Paraoxon

Humans lack this enzyme. The toxicity of parathion in insects and humans is given below.

Compound:	Parathion	Paraoxon
Structure:	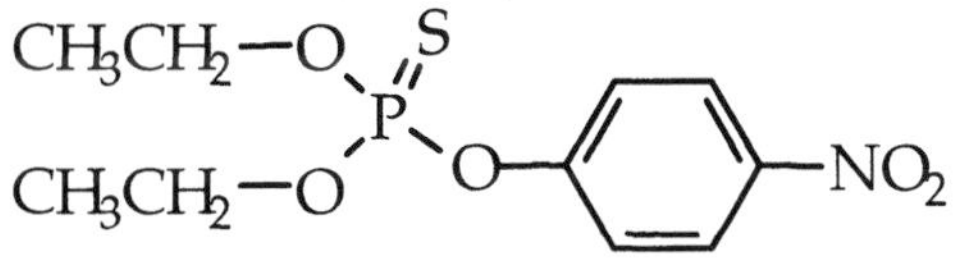	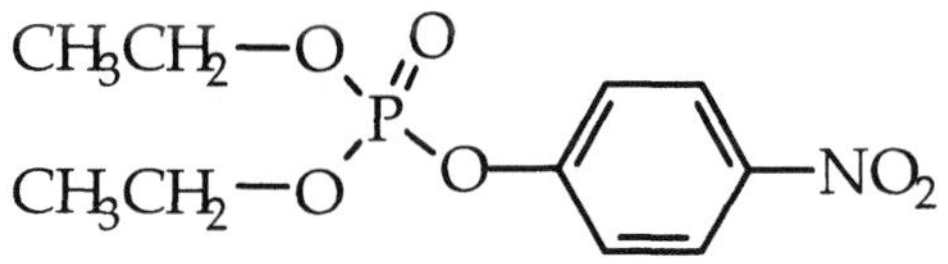
Human Toxicity:	low	high
Insect Toxicity:	high	high

Explain how exposure to parathion can be more toxic to insects than humans.

(3.1.7) A class of drugs called nonsteroidal anti-inflammatory drugs (NSAIDs for short) are used to reduce pain and inflammation. Aspirin was the first drug of this group; it was developed more than 100 years ago. Aspirin is a very effective pain killer, but long-term use can lead to stomach ulcers. A stomach ulcer is where the lining of cells in the stomach is breached.

Aspirin acts by inhibiting two enzymes, cyclo-oxygenase 1 and cyclo-oxygenase 2 (COX-1 and COX-2). COX-1 and COX-2 have very similar structures. These enzymes are involved in the following pathway (you may want to look up prostaglandins in your textbook):

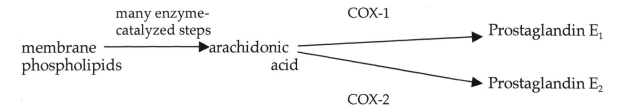

Membrane phospholipids are converted to arachidonic acid by a series of enzyme-catalyzed reactions. Arachidonic acid is converted to Prostaglandin E_1 by COX-1; it is converted to Prostaglandin E_2 by COX-2. In the absence of any drugs, arachidonic acid is converted to both prostaglandins. The prostaglandins are hormones that circulate in the blood. They have very different effects in the body:

Prostaglandin E_1 is required to maintain the layer of cells that line the stomach. Without Prostaglandin E_1, these cells are not replaced as they die.

Prostaglandin E_2 binds to receptors on the surface of pain-sensitive cells and increases their sensitivity to pain. That is, in the presence of Prostaglandin E_2, pain-sensitive cells send stronger and more frequent pain messages to the brain.

a) Aspirin inhibits both COX-1 and COX-2 by covalently binding to the -OH of a serine at the active site of both enzymes.
　　　i) Based on this, why does aspirin act as an analgesic (pain reliever)?

　　　ii) Based on this, why does extended use of aspirin lead to stomach ulcers?

b) Newer NSAIDs, like Celebrex and its relatives, are designed to inhibit COX-2 only. Why would Celebrex and other COX-2 inhibitors be better pain relievers than aspirin?

(3.2) Biochemical pathways, general

Diagnostic Problem:

In an organism where:

$$\Rightarrow \text{compound } \alpha \quad \overset{\text{enzyme 1}}{\Rightarrow} \quad \text{compound } \beta \quad \overset{\text{enzyme 2}}{\Rightarrow} \quad \text{compound } \gamma \quad \overset{\text{enzyme 3}}{\Rightarrow} \quad \text{final product,}$$

you could assume that an organism lacking only enzyme 1 would not form β and could not therefore make the final product. However, if this organism had an exogenous supply of β, enzyme 2 could convert β into γ and enzyme 3 could convert γ into the final product.

a) What compound or compounds could be made if an organism lacked both enzyme 1 and enzyme 2?

b) If an organism that lacked both enzyme 1 and enzyme 2 had an exogenous supply of β, would it be able to make the final product?

c) Given the pathway above, an organism lacking enzyme 2 might accumulate which compound?

d) Given the pathway above, an organism lacking both enzyme 1 and enzyme 2 might accumulate which compound?

Solutions to Diagnostic Problem:

a) What compound or compounds could be made if an organism lacked both enzyme 1 and enzyme 2?
Only compound α could be made.

b) If an organism that lacked both enzyme 1 and enzyme 2 had an exogenous supply of β, would it be able to make the final product?
Even if this organism had an exogenous supply of β it could not make γ, so it could not make the final product. Indeed, the organism lacking both enzyme 1 and enzyme 2 would need an exogenous supply of γ to complete the pathway and make the final product.

c) Given the pathway above, an organism lacking enzyme 2 might accumulate which compound?
An organism lacking enzyme 2 would accumulate β.

d) Given the pathway above, an organism lacking both enzyme 1 and enzyme 2 might accumulate which compound?
An organism lacking both enzyme 1 and enzyme 2 would accumulate α.

(3.2.1) The following represents a pathway for the synthesis of the essential compound A in a bacterial cell.

$$\text{compound X} \overset{\text{enzyme 1}}{\Rightarrow} \text{compound Y} \overset{\text{enzyme 2}}{\Rightarrow} \text{compound Z} \overset{\text{enzyme 3}}{\Rightarrow} \text{compound A}$$

Bacterial cells with defective enzyme 1, 2, or 3 will grow only if compound A is available (added to the growth media).

a) What compound will build up in the cells defective in the following enzymes?

enzyme 1:   enzyme 2:   enzyme 3:

b) What compound(s) must be available to cells defective in the following enzymes?

enzyme 1:   enzyme 2:   enzyme 3:

c) What compound will build up in the cells defective in the following enzymes?

enzyme 1 and enzyme 2:   enzyme 2 and enzyme 3:   enzyme 1 and enzyme 3:

d) What compound(s) must be available to cells defective in the following enzymes?

enzyme 1 and enzyme 2:   enzyme 2 and enzyme 3:   enzyme 1 and enzyme 3:

(3.2.2) You are studying the biosynthesis of the amino acid arginine. You have cells that are missing different enzymes needed in this pathway. You can add three potential compounds (A, B, C) as a supplement to the medium. You test your cells to determine whether they grow (+) or not (−).

Type of cell	Medium				
	Nothing added	Compound A added	Compound B added	Compound C added	Arginine added
Normal	+	+	+	+	+
Missing enzyme 1	−	−	+	−	+
Missing enzyme 2	−	−	−	−	+
Missing enzyme 3	−	+	+	−	+
Missing enzyme 4	−	+	+	+	+

What is the order of enzymes and compounds in the pathway?

(3.2.3) In certain bacteria, the biosynthesis of aromatic amino acids (phenylalanine, Phe; tyrosine, Tyr; and tryptophan, Trp) proceeds by a branched pathway where some intermediates are used to make more than one amino acid. The pathway and some key intermediates are shown below.

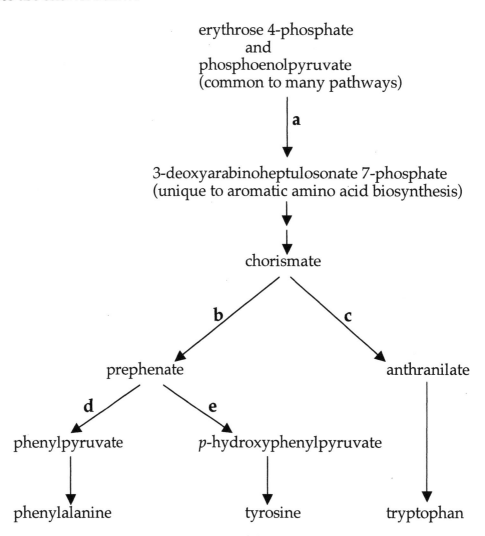

This pathway is regulated by feedback inhibition at enzymes **a** through **e**. For enzymes **a** through **e**, list the intermediate or product shown that you expect to be the <u>most likely inhibitor</u> of each enzyme (for a description of the different types of inhibition, see your textbook). Provide a brief explanation for each answer.

(3.3) Glycolysis, respiration, and photosynthesis

(3.3.1) There are only a few different classes of enzymes that catalyze the reactions of the glycolytic pathway.

a) Name <u>two</u> enzymatic steps in glycolysis that are similar to the following reaction. Write the names of the reactant, the product, and the enzyme used in each of these two reactions.

$$\text{ATP} + \underset{\text{CH}_3}{\overset{\overset{\textstyle CH_3}{|}}{CH_3C\text{-}OH}} \longleftrightarrow \underset{\text{CH}_3}{\overset{\overset{\textstyle CH_3}{|}}{CH_3C\text{-}O}}\overset{\overset{\textstyle O}{||}}{\underset{\underset{\textstyle O\text{-}}{|}}{P}}\text{-}O\text{-} + \text{ADP}$$

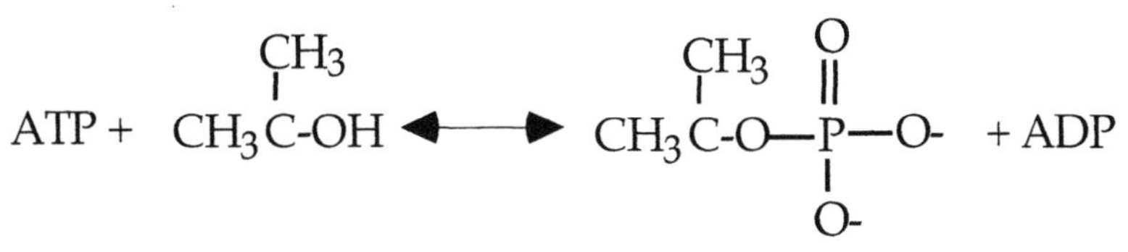

b) Name <u>two</u> enzymatic steps in glycolysis that are similar to the following reaction. Write the names of the reactant, the product, and the enzyme used in each of these two reactions.

$$\begin{array}{c} CH_2OH \\ | \\ C{=}O \\ | \\ CH_2CH_3 \end{array} \longleftrightarrow \begin{array}{c} O{\diagup}^{\diagdown}H \\ C \\ | \\ H\text{-}C\text{-}OH \\ | \\ CH_2CH_3 \end{array}$$

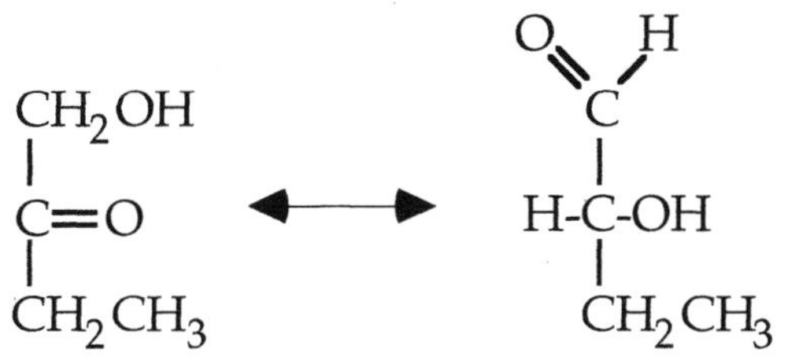

c) If you add each of the above substrates (shown above) to the corresponding glycolytic enzymes you indicated above, you will find that these substrates will not necessarily be converted into the given products. Why do you think this is the case?

d) One of the reactions of the citric acid cycle (Krebs cycle) is catalyzed by succinate dehydrogenase. This step is inhibited by adding malonate, shown below, to the solution.

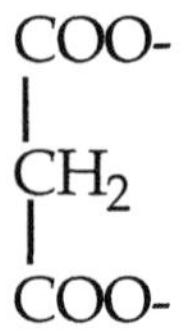

$$
\begin{array}{c}
\text{COO-} \\
| \\
\text{CH}_2 \\
| \\
\text{COO-}
\end{array}
$$

Malonate

Explain why malonate acts as a competitive inhibitor of the succinate dehydrogenase enzyme.

(3.3.2) In the 1860s, Louis Pasteur observed the following phenomenon, which has come to be called "the Pasteur Effect." If you grow a culture of *Escherichia coli* bacteria (which can grow anaerobically or aerobically) without O_2, they consume large amounts of glucose as they grow and they produce lactic acid from the glucose. If you now supply this culture with an excess of O_2, two things happen rapidly.

  (1) Lactic acid is no longer produced.
  (2) The rate of glucose consumption decreases even though the rate of cell growth is constant.

a) Explain (1) in terms of the biochemical fate of glucose and its derivatives; what happens to the glucose if it isn't converted to lactic acid? (You need not list the entire pathway, just the key intermediates and end products.)

b) How is it that less glucose is required for the same growth rate once O_2 is added?

c) Small amounts of NAD^+ are sufficient to metabolize large amounts of glucose. Explain why this statement is true under both anaerobic and aerobic conditions.

(3.3.3) The first step in the pathway of glycolysis (see your textbook) is the transfer of a phosphate group from ATP to glucose, giving glucose 6-phosphate (reactions of this type are phosphorylations). The next step in glycolysis is the isomerization of glucose 6-phosphate to fructose 6-phosphate.

Cells can also use fructose as an energy source. Through a pathway similar to early steps in glycolysis, fructose is converted to an intermediate in the glycolytic pathway (i.e., it enters glycolysis at a middle step). In the first step of fructolysis, fructose is phosphorylated to convert it to fructose 1-phosphate. The phosphorylation of fructose is catalyzed by the enzyme fructokinase. This reaction is shown below:

Reaction 1: fructose + ATP $\Rightarrow$ fructose 1-phosphate + ADP

a) Fructose 1-phosphate is then metabolized to glyceraldehyde and dihydroxyacetone phosphate (DHAP) by the enzyme fructose-1-phosphate aldolase. DHAP enters glycolysis directly. However, the glyceraldehyde must first be phosphorylated to glyceraldehyde 3-phosphate by triose kinase. Which molecule would most likely be used to provide the phosphate group and the energy required to phosphorylate glyceraldehyde? Write the chemical reaction for this phosphorylation. (You need not draw structures, just use chemical names.)

b) Based on the known glycolytic pathway (see your textbook), reaction 1, and the reaction you postulated in part (a), draw the pathway from fructose to two molecules of pyruvate. (You need not draw chemical structures, just use chemical names.)

c) Based on the pathway you drew for part (b), how much ATP is consumed in the conversion of fructose into two pyruvate molecules? How much ATP is generated? How does the net result of converting fructose to two molecules of pyruvate compare with the conversion of glucose to two molecules of pyruvate?

(3.3.4) During fermentation, glucose is broken down into pyruvate via glycolysis (see your textbook); the pyruvate is then converted to CO_2 and ethanol (see your textbook). This is called the Embden-Meyerhoff-Parnas pathway for the researchers who worked it out. The stoichiometry of the overall reaction is:

$$\text{glucose} + 2\text{ ADP} + 2\text{ P}_i \Rightarrow 2\text{ CO}_2 + 2\text{ ethanol} + 2\text{ ATP} + 2\text{ H}_2\text{O}$$

There is an alternative fermentation pathway for the anaerobic breakdown of glucose: the Entner-Doudoroff pathway. What follows is a list, in random order, of the enzymes of the Entner-Doudoroff pathway and the reactions they catalyze (note that the longer chemical names have abbreviations shown in parentheses):

Notes:
- For the overall reaction: the reactants are glucose, ADP, and P_i; the products are CO_2, ethanol, ATP, and H_2O.
- No NAD^+ or NADH are produced or consumed by the <u>overall</u> reaction. That is, any NAD^+ or NADH produced by one reaction must be recycled by another reaction.

<u>Enzyme</u>	<u>Reaction Catalyzed</u>
A	6-phosphogluconate (6PG) $\longrightarrow$ 2-keto-3-deoxy-6-phosphogluconate (2,3,6PG) + H_2O
B	glucose 6-phosphate (G6P) + H_2O + NAD^+ $\longrightarrow$ 6-phosphogluconate (6PG) + NADH + H^+
C	1,3-bisphosphoglycerate (BPG) + ADP $\longrightarrow$ 3-phosphoglycerate (3PG) + ATP
D	3-phosphoglycerate (3PG) $\longrightarrow$ 2-phosphoglycerate (2PG)
E	glyceraldehyde 3-phosphate (G3P) + P_i + NAD^+ $\longrightarrow$ 1,3-bisphosphoglycerate (BPG) + NADH + H^+
F	2-phosphoglycerate (2PG) $\longrightarrow$ phosphoenolpyruvate (PEP) + H_2O
G	glucose + ATP $\longrightarrow$ glucose 6-phosphate (G6P) + ADP
H	phosphoenolpyruvate (PEP) + ADP $\longrightarrow$ pyruvate + ATP
I	acetaldehyde + NADH + H^+ $\longrightarrow$ ethanol + NAD^+
J	pyruvate $\longrightarrow$ acetaldehyde + CO_2
K	2-keto-3-deoxy-6-phosphogluconate (2,3,6PG) $\longrightarrow$ pyruvate + glyceraldehyde 3-phosphate (G3P)

a) Draw the Entner-Doudoroff pathway, indicating the order of reactions, the enzyme that catalyzes each step, and the intermediates. Your diagram should look like the figure in your textbook except that you can use names or abbreviations for the intermediates; be sure to label each reaction with the letter that corresponds to the enzyme that catalyzes it.

b) What is the stoichiometry of the overall reaction of the Entner-Doudoroff pathway? How many ATPs does a bacterium using this pathway get per glucose consumed; how does this compare with the ATP yield from the Embden-Meyerhoff-Parnas pathway?

c) The two pathways have several identical enzymatic steps. For each step in the Entner-Doudoroff pathway (A through K), either:

   (1) Give the identical step in the Embden-Meyerhoff-Parnas pathway; give the figure from your textbook where the corresponding step appears and the step number in that figure.

or: (2) State that there is no equivalent step.

d) The two pathways described above consume hexoses: 6-carbon sugars like glucose. Bacteria can also grow on pentoses (5-carbon sugars) like xylose. First, the xylose must be processed into a usable form by the pentose phosphate pathway. The overall reaction of the pentose phosphate pathway is:

3 xylose + 3 ATP $\Rightarrow$ 2 glucose 6-phosphate (G6P) + glyceraldehyde 3-phosphate (G3P)
+ 3 ADP

Note that carbon atoms are conserved in these reactions: $3 C_5 \Rightarrow 2 C_6 + C_3$.

What would be the stoichiometry of the overall reaction if you combined the pentose phosphate pathway with the Entner-Doudoroff pathway to convert xylose, ADP, and P_i to ethanol, H_2O, ATP, and CO_2?

e) How many ATPs does a bacterium using these combined pathways get per xylose consumed?

(3.3.5) When exposed to light, plant cells show net absorption of CO_2 and net production of O_2. In the dark, they show net production of CO_2 and net absorption of O_2.

a) What biochemical process is responsible for the plant's absorption of O_2 and production of CO_2 in the dark?

b) Does this process continue when the plant is exposed to light? If so, why aren't net production of CO_2 and absorption of O_2 seen under these conditions? Explain briefly.

Chapter 2: Biochemistry Problems

(3.3.6) Refer to the figure in your textbook that shows photosynthetic electron transport. The Hill reagent is an organic molecule that can bind to NADP-reductase and compete with $NADP^+$ as an electron acceptor. However, once it picks up electrons, the Hill reagent cannot transfer them to any biomolecules. When large amounts of the Hill reagent are added to plant cells, all the electrons from noncyclic photophosphorylation are transferred to the Hill reagent instead of to $NADP^+$. Under these conditions and in the presence of CO_2, H_2O, and light, O_2 production continues for a short time, but eventually stops.

a) How can O_2 continue to be produced by these cells for a short time? Explain your answer briefly.

b) Why does O_2 production stop eventually? Explain your answer briefly.

c) The herbicide DCMU {[3-(3,4-dichlorophenyl)-1,1-dimethylurea]} binds to plastoquinone (also called pQ) and inactivates it so that electrons can no longer be transferred through pQ.

   i) Will O_2 be produced by these cells? Explain your answer.

   ii) Will ADP + P_i still be converted to ATP by the chloroplasts of these cells? Explain your answer briefly.

   iii) Will CO_2 still be reduced to glucose? Explain your answer briefly.

(3.3.7) It is now possible to determine the DNA sequence of the entire genome of an organism. This allows you to make a list of all the proteins that this organism can produce. From this, it is sometimes possible to deduce an organism's metabolic behavior.

This has been done for a number of bacteria. This problem deals with four:
- *Haemophilus influenzae*, a human pathogen that causes meningitis, among other nasty diseases.
- *Mycoplasma genitalium*, a nonpathogenic bacterium.
- *Mycoplasma pneumoniae*, which causes some cases of pneumonia.
- *Escherichia coli*, a usually harmless bacterium.

This problem also deals with six enzymes that we will use as markers for the pathways they participate in. They are:

Enzyme	Reaction Catalyzed
lactate dehydrogenase (lactate DHase)	pyruvate + NADH $\Rightarrow$ lactate + NAD$^+$
alcohol dehydrogenase (Alcohol DHase)	acetaldehyde + NADH $\Rightarrow$ ethanol + NAD$^+$
aldolase	fructose 1,6-bisphosphate $\Rightarrow$ dihydroxyacetone phosphate + glyceraldehyde phosphate
proton ATPase	lets H$^+$ move through membrane to produce ATP from ADP + P$_i$
cytochrome c oxidase	transfers electrons from cytochrome c_1 to cytochrome a
citrate synthase	oxaloacetate + acetyl-CoA --> citrate

a) For each of the enzymes listed above, determine the part of energy metabolism (glycolysis, fermentation, electron transport, etc.) that the enzyme belongs to. You will have to look through the sections of your textbook that deal with glycolysis, fermentation, electron transport, and oxidative phosphorylation. Some of the names may be slightly different in different texts.

In general, if one enzyme in a part of energy metabolism is present, then all the related enzymes are also present. In this way, the presence of each of the enzymes above can be used as a marker for the rest of that part of energy metabolism. Using this, you can figure out the kinds of energy metabolism that an organism can carry out.

The table below shows the presence (+) or absence (–) of each of the enzymes above in the genomes of the bacteria in this problem:

Organism	Lactate DHase	Alcohol DHase	Aldolase	Proton ATPase	Cytochrome c oxidase	Citrate synthase
Haemophilus influenzae	–	+	+	+	–	–
Mycoplasma genitalium	+	–	+	+	–	–
Mycoplasma pneumoniae	+	–	+	+	+	–
Escherichia coli	–	+	+	+	+	+

b) For each of the four bacteria, answer the following questions based on the data above:

- If this organism is grown on glucose in the **presence** of oxygen, will $CO_2 + H_2O$, alcohol + CO_2, or lactic acid be produced? Roughly how many ATPs will the cell get from each molecule of glucose under these conditions?

- If this organism is grown on glucose in the **absence** of oxygen, will CO_2 only, alcohol + CO_2, or lactic acid be produced? Roughly how many ATPs will the cell get from each molecule of glucose under these conditions?

Chapter 3:

Molecular Biology Problems

Molecular Biology Problems

If you were a molecular biologist, you would focus on biological molecules like DNA, RNA, and proteins. Although generally true, your work would overlap with other areas like genetics and biochemistry. In this chapter, we have given you problems that will allow you to explore the structure and function of DNA and RNA and how proteins are elaborated in the cell.

The first problems examine some of the most important experiments that led to the conclusion that DNA is the genetic material. They are good examples of the history of science as well as opportunities to analyze real data. Before attempting the question in this section, refer to your textbook or lecture material.

The first diagnostic question for this chapter is found in section 2. Work through the diagnostic question on your own and then look at our approach to solving it. If any of the terms are unfamiliar, consult the appropriate chapter in your textbook.

(1) PROBLEMS EXPLORING CLASSIC EXPERIMENTS
Key words: "Griffith," "*Streptococcus pneumoniae*," "Transforming Principle," "Avery," "Pneumococcus," "Hershey," and "Chase."

Note that this section does not have a diagnostic problem; you should consult your textbook for further information if you have difficulty working these problems.

(1.1) Frederick Griffith showed that a component of dead bacterial cells could confer new properties to live bacteria of the same species. The property in question was the presence of a surface polysaccharide capsule of the bacterium *Streptococcus pneumoniae*, the bacterium that causes pneumonia.

Two types of these bacteria are found, based on two main types of surface polysaccharide:
> 1) R (rough) cells have no polysaccharide capsule. These bacteria do not cause disease. They are nonvirulent.
> 2) S (smooth) cells have a polysaccharide capsule. These bacteria cause disease and are virulent.

The S (smooth) cells fall into several different subtypes based upon the polysaccharide capsule. These are designated S_I through S_{XXIII}.

You can isolate mutants of all these S strains that no longer make a capsule and are no longer virulent. The lack of a polysaccharide capsule makes one R strain indistinguishable from another. However, R strains are also designated $R_{(I)}$ through $R_{(XXIII)}$, depending upon their origin. An R strain derived from S_{III} is designated $R_{(III)}$.

The central experiment was:
Experiment: Inject a mixture of $R_{(II)}$ and heat-killed S_{III} into a mouse.
Results: The mouse dies of pneumonia, and virulent S_{III} bacteria can be isolated from the mouse.

Griffith et al. also did some control experiments to support experiment 1.
Negative Control: Inject $R_{(II)}$ or heat-killed S_{III} **alone** into a mouse.
Results: The mouse lives.

Positive Control: Inject heat-killed S_{III} alone into a mouse.
Results: The mouse dies of pneumonia, and only virulent S_{III} bacteria can be isolated from the mouse.

a) For each of the control experiments listed, explain:

- What alternative models does the result of this control experiment rule out? That is, complete the sentence, "Without this control result, you could argue that the mice died in experiment 1 because...."

- What would it have meant for their model if the results of this control experiment were the reverse of those expected (the mouse dies instead of living and vice versa), assuming that the results of experiment 1 were the same as above?

b) Another potential problem with their experiments was the possibility that, at a low frequency, $R_{(II)}$ can mutate back to the S_{II} from which it was derived. (Note: $R_{(II)}$ cannot revert to any other subtype of S.) Based on this, why was it essential that they use a mixture of $R_{(II)}$ and heat-killed S_{III} instead of a mixture of $R_{(II)}$ and heat-killed S_{II}?

(1.2) The team of Avery, McCarty, and MacLeod was attempting to purify a substance from smooth (S) pneumococci which was capable of transforming rough (R) pneumococci into smooth; they called this substance the "transforming substance."

a) They presumed that the transforming substance was genetic material (a.k.a. genes). Explain why they believed that this was so.

The authors were trying to distinguish between two models for the "transforming substance" (genes):
    (1) Genes are made of protein.
    (2) Genes are made of DNA.

It was their belief that model (2) was correct. They wanted to purify the transforming substance away from other cellular material and then determine whether it was pure DNA or a mixture of protein and DNA. They hoped to show that protein was either not present or not essential for the transforming substance to be able to transform R to S.

b) If they had found traces of protein in their preparations of the transforming substance (even if it was >99% DNA), this would have made it impossible to rule out model (1). Explain why this is so.

c) In their paper, they presented an analysis of the elemental composition of several of their purified preparations. In particular, they compared the ratio of nitrogen to phosphorus (N/P ratio) of their preparations, with the N/P ratio predicted based on the structure of DNA.

    i) Using the structure of DNA from your text, the atomic weight of N = 15 and the atomic weight of P = 31, show that the ratio for the DNA:

$$\frac{\text{total mass of N}}{\text{total mass of P}} = 1.69.$$

    ii) If their preparations were contaminated with protein, would you expect the ratio of the preparations to be higher or lower than 1.69? Explain your reasoning.

These are their actual data (all of these preparations could transform R to S):

Preparation #	N/P ratio
37	1.66
38B	1.75
42	1.69
44	1.58

   iii) What conclusions would you draw from these data, and what would be your reservations about these conclusions?

d) In another series of experiments, preparations were treated with various enzymes of known function. They wanted to determine whether these enzymes were capable of destroying the transforming ability of the preparations. They treated their preparations with the following enzymes alone or in combination:
   • Trypsin: an enzyme that breaks proteins down into amino acids.
   • Chymotrypsin: another enzyme that breaks proteins down into amino acids.
None of these treatments had any effect on the transforming ability of their preparations.

   i) Why do these data support model (2)?

   ii) Why are these data, on their own, not completely conclusive? If you believed in model (1), how would you argue that these data are consistent with model (1)?

(1.3) Hershey and Chase provided strong evidence that DNA, and not protein, is the genetic coding material of the cell through their experiments involving differential partitioning of ^{32}P-labeled DNA and ^{35}S-labeled protein of bacteriophage T2.

a) In their studies, there were two crucial experiments.
Experiment 1: Bacteriophage were labeled with ^{32}P, which is incorporated only into DNA.
Experiment 2: Bacteriophage were labeled with ^{35}S, which is incorporated only into protein.

The bacteriophage were allowed to attach to the bacterial cells and transmit the genetic material. The bacteria and the bacteriophage were then separated, and the amount of radioisotope in each was measured. For each scenario below, describe how much radioactivity you would expect to find in the phage as opposed to the bacteria (for example: a lot, some of it, little, or none). Explain your reasoning briefly.

i) If the phage contained both DNA and protein but the genetic material injected was protein and the DNA remains in the phage head.

ii) If the genetic material were a mixture of mostly DNA and a little protein.

iii) If the genetic material were protein that is carried on the DNA and the DNA is only a scaffold for carrying the protein genetic material.

b) Actually, the data were not as unambiguous as described in most textbooks. (For an interesting discussion of how this experiment has changed in the telling and retelling, see "How history has blended," *Nature* **249**:803-805, 1974.) They performed the experiment several times and got the following results:
   (1) Between 15% and 35% of the ^{32}P was found in the supernatant.
   (2) Between 18% and 25% of the ^{35}S was found in the pellet.
Assume that their model is correct (the phage injects DNA and not protein).

i) What could have caused the presence of the ^{32}P in the supernatant? How crucial for their model is it that this number be 0? Why?

ii) What could have caused the presence of some of the ^{35}S in the pellet? How crucial for their model is it that this number be 0%? Why?

c) Based on the above data, which of the models (i, ii, or iii) in part (a) can you rule out? Explain your reasoning.

(1.4) On a mission to a new solar system, you discover an alien virus that contains nucleic acids, proteins, and lipids. You also find that this virus can infect *E. coli* cells, making it easy to study in the laboratory.

a) You grow this virus with one of the following radioisotopes: ^{32}P, ^{3}H, or ^{35}S.

- Which of the viral macromolecules will be labeled with ^{32}P?

- Which of the viral macromolecules will be labeled with ^{3}H?

- Which of the viral macromolecules will be labeled with ^{35}S?

b) You analyze the nucleic acid and find the following:

Percentage of each base:

A	G	T	C	U
27.6	23.2	28.1	24.0	0.8

What nucleic acid is the virus carrying? How do you know this?

c) Because this is an alien virus, you want to determine which of the macromolecules (nucleic acids, proteins, and lipids) is the hereditary material. Explain how you would do this.

d) You examine DNA replication to determine whether it is similar to DNA replication on Earth. You begin by constructing an in vitro system for DNA replication.

    i) Assuming that the process of alien DNA replication is similar to that seen on Earth, what four components would you include in your system?

    ii) Prior to replication, all the template DNA is labeled with ^{15}N. Where is nitrogen found in DNA?

    iii) You repeat the Meselson-Stahl experiments. On the diagram below, draw the results expected at each round for both conservative and semiconservative replication.

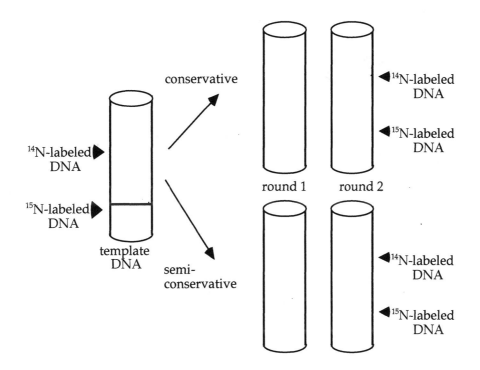

(2) PROBLEMS EXPLORING THE STRUCTURE OF DNA AND RNA

Although biology is often a science of exceptions, there are several important concepts that apply to the nucleic acids, DNA and RNA. These statements hold in the vast majority of cases.

Directionality – All strands have a direction. This is specified in terms of the 5′ and 3′ carbons in the sugar backbone.

Antiparallel – In order to form proper base pairs, the two strands must run 5′ to 3′ in opposite directions.

Base pairing – In DNA, A always pairs only with T. In RNA, A pairs with U. G always pairs only with C.

Polymerization – New nucleotides are always added only to the 3′ end of a chain. Thus, polymerization always occurs in a 5′ to 3′ direction.

DNA and RNA Structure:

1) Building blocks of DNA are nucleotides:

Base =
Adenine (A) } purines
Guanine (G) }
Cytosine (C) } pyrimidines
Thymine (T) }

2) The nucleotides in DNA are joined by 3′ – 5′ phosphodiester bonds.

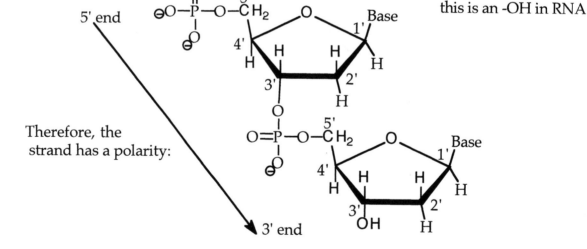

this is an -OH in RNA

5′ end

Therefore, the strand has a polarity:

3′ end

3) The DNA has two strands held together by hydrogen bonds between the bases on opposite strands.

4) The two strands have opposite polarity.
5) The two strands are twisted into a right-handed double-helix.
6) RNA is single-stranded, but it can fold on itself to create regions where hydrogen bonds form between the bases.

Diagnostic Question:

You are given the sequence of one strand of a DNA molecule:

5′ end
→ A A T C G G C T T A C C T A C C A T T T T A ← 3′ end

a) Directly below the sequence above, give the sequence of the second strand of DNA. Label all 3′ and 5′ ends.

b) What chemical group is usually found on the 3′ end?

c) What chemical group is usually found on the 5′ end?

d) What holds the two strands to each other?

e) A new strand of DNA is made in what direction?

Answer to Diagnostic Question:

a)

5′ end   3′ end
Original strand: → A A T C G G C T T A C C T A C C A T T T T A ←

New strand: → *T T A G C C G A A T G G A T G G T A A A A T* ←
              3′ end   5′ end

b) What chemical group is usually found on the 3′ end? *There is usually a hydroxyl at the 3′ end.*

c) What chemical group is usually found on the 5′ end? *There is usually a phosphate group at the 5′ end.*

d) What holds the two strands to each other? *Hydrogen bonds hold the two strands together.*

e) A new strand of DNA is made in what direction? *A new strand of DNA is made in the 5′ to 3′ direction.*

(C1) Computer Activity 1. This activity will guide you through the three-dimensional structure of DNA.

Open the "Biochemistry" folder on the CD-ROM and double-click on the "Molecules in 3-dimensions" program to launch it. Note that Molecules in 3-dimensions can sometimes take a while to get started.

You will see the following on a PC:

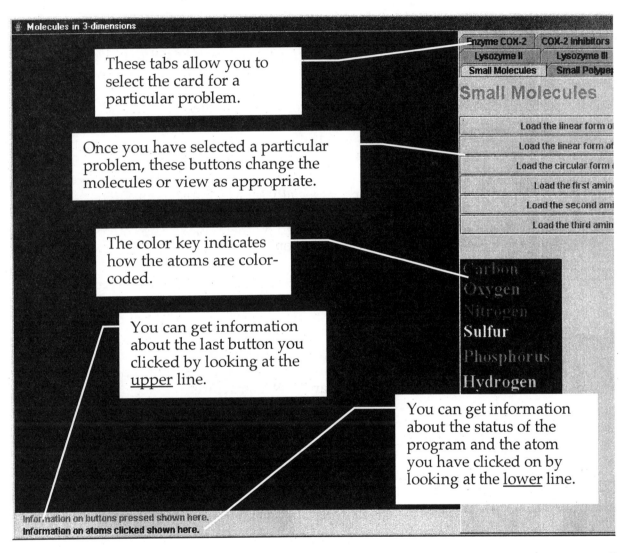

On a Macintosh, the problem selection tab looks like this (the rest is very similar to the screen shown above). You can see the tabs for two problems; you access the reminder by clicking the arrow.

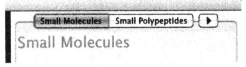

This software has many useful features, so we will take some time now to describe its use in detail. This software allows you to get information from the image in several ways:

- <u>Rotating the molecule</u>: This is the best way to get an idea of the molecule's three-dimensional structure. You can click and drag on any part of the molecule and it will rotate as though you had grabbed it.
- <u>Zooming in or out</u>: This helps to get close-up or "big-picture" views of the molecule. Hold the shift key down while dragging the cursor up (to zoom out) or down (to zoom in) the image.
- <u>Identifying the atom you are looking at</u>: You can find information on the atoms in the molecule in one of two ways:
  - By clicking on an atom and looking at the lower left of the "Molecules in 3-dimensions" window. A small line of text will appear there with information on the atom you just clicked.
  - By putting the cursor over the atom you are interested in and waiting a few seconds for the information to pop up. The program will then display information on the atom in a little pop-up window. The information in the pop-up window is more detailed than the above but rather cryptic. Try putting the cursor over some other atoms to see what you get. Note that this does not always work, especially on Macintosh computers.

In addition to the above, atoms are also identified by their color. The color scheme is shown to the right of the molecule images.

Here are a few important notes about the way these molecules are displayed in this problem:

- The atoms are colored according to the scheme at the right of the window.
- Only covalent bonds are shown.
- All covalent bonds are shown as single lines; single, double, and triple bonds are all shown identically. You have to figure out the bond type based on your knowledge of covalent bonding and DNA structure.
- Hydrogen atoms are not shown in these structures. This is because these are actual protein structures determined by X-ray crystallography. Hydrogen atoms are not visible in X-ray crystallograms and are therefore not shown in these structures. You have to infer the hydrogens yourself based on your knowledge of covalent bonds and DNA structures.

Click the tab for this problem: "DNA Structure."

a) Click the button marked "Load First DNA Molecule." You will see a black window with a DNA molecule shown in "spacefill" mode where atoms are shown as solid spheres at their actual sizes. You can click on the "Show atoms as ball and stick" button to change the representation to "ball and stick" where atoms are shown as balls and the covalent bonds connecting them are shown as rods. You will find it useful to switch back and forth between the two views. Note that you may sometimes need to click this button three times to get the view to change.
- The ball and stick view shows covalent bonds as rods and is most useful for determining which atoms are *covalently bonded* to each other.
- The spacefill view shows atoms as joined spheres of their approximate actual size in the molecule. It is most useful for determining which atoms are *closest together*. Because it does not show covalent bonds, it can sometimes be hard to figure out which atoms are covalently bonded.

i) Click on the "Show the two strands" button and select the "spacefill" view. In this view, the atoms are colored as follows:
- One sugar-phosphate **backbone** strand is **pink**.
- The other sugar-phosphate **backbone** strand is **light yellow**.
- The **5' carbons** at the ends of the two backbone strands are **purple**.
- The **3' carbons** at the ends of the two backbone strands are **white**.
- The **bases** are **green**.

The two sugar-phosphate backbone strands run next to each other. Based on the structure, are they parallel (both run 5' to 3' in the same direction) or antiparallel (run 5' to 3' in opposite directions)?

ii) Switch to a "ball and stick" view. Based on this view, are there any covalent bonds between the bases on the two different strands?

iii) Click on the "Color-code bases and the two strands" button. This is the same as (i) and (ii), except the bases are colored to identify which type they are.
- **Adenine** is **yellow**.
- **Thymine** is **red**.
- **Cytosine** is **blue**.
- **Guanine** is **green**.

Based on this view, which bases pair with which? By clicking on individual atoms, you can identify the different bases as you click along the chain. Using this technique, determine the sequence of as much of both DNA strands as you can.

b) Click the "Load Second DNA Molecule" button. You will see a short (3-base-pair) double-stranded DNA. This illustrates the details of DNA structure. You should switch to the "ball and stick" view to help you see this more clearly. The atoms are colored using the color scheme on the right of the window **except**:

- **5′ carbons** are **purple**.
- **3′ carbons** are **white**.

i) Look at the 5′ and 3′ carbons and determine the 5′ to 3′ direction of each strand. Note that every nucleotide contains both a 5′ and a 3′ carbon. If you look at the strands, you will see that, in one strand, the 5′ carbons in each sugar come first, while in the other it is reversed.

ii) Determine the sequence of both strands of the short DNA molecule shown. The "spacefill" view may be best for this. Hint: you don't have to look at every atom in the bases: the bases are colored as described. Also, you can click on atoms in each base and they will be identified in the information line at the bottom of the window.

(2.2) Consider the following DNA segment that has begun to unwind for replication. The arrow represents the first nucleotides of the newly formed DNA. The arrow point marks the site where a new nucleotide will be added.

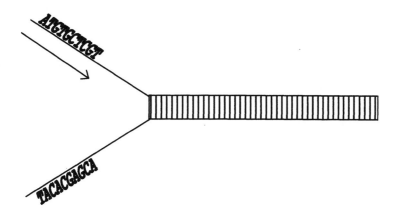

a) What is the sequence of the newly formed DNA?

b) Label all 5′ and 3′ ends.

c) Draw an arrow on the bottom strand such that the arrow point marks the site where a new nucleotide will be added.

(2.3) Look at the following schematics of RNA molecules. The hydrogen bonds between the bases are indicated by the lines.

a) Given that base pairing requires the strands to be antiparallel, circle the RNA molecules that could form.

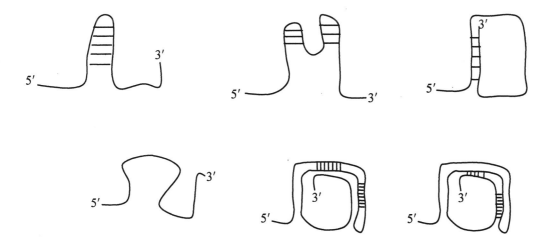

b) If the sequence in a region were 5′… ACGGACGC…3′, what would be the sequence of the DNA that is hydrogen bonded to it?

(2.4) The next nucleotide to be added to a growing DNA strand is dCTP (shown).

- Circle the part of the growing DNA chain to which the next base is attached.
- Circle the part of the dCTP that is incorporated into the growing DNA chain.

(3) DNA REPLICATION

Diagnostic Question:

Shown below are two sequential close-up views of the same replication fork.

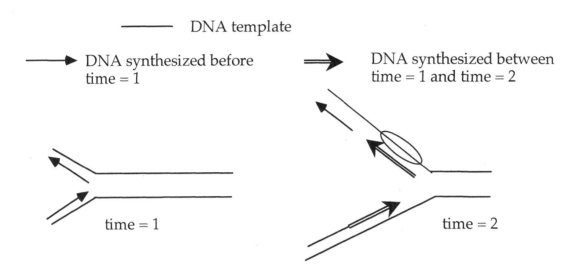

Arrows indicate the direction of DNA polymerization.

a) Label the 5' and 3' ends of all the DNA molecules shown.

b) The sequence within the circled region on the top template strand was as follows:
$$5' \ldots ATTCCG \ldots 3'$$

- Give the sequence of the complementary DNA.

- Box the area where this complementary sequence is found.

c) As the two template strands are pulled apart, synthesis of new DNA using the bottom template strand is continuous. Why is synthesis of new DNA using the top strand discontinuous?

d) Label the leading strand and the lagging strand.

Answer to Diagnostic Question:

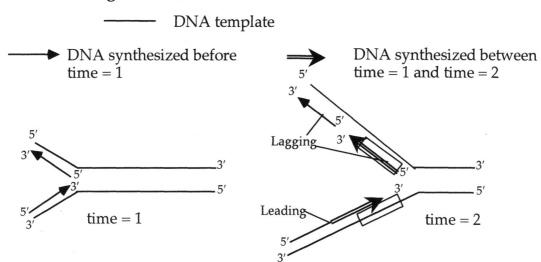

Arrows indicate the direction of DNA polymerization.

The answer to this question is really a restatement of some of the concepts given at the beginning of this section.

- **Directionality** – All DNA strands have a direction. This is specified in terms of the 5′ and 3′ carbons in the sugar backbone.
- **Antiparallel** – In order to form proper base pairs, the two strands must run 5′ to 3′ in opposite directions.
- **Base pairing** – A always pairs only with T. G always pairs only with C.
- **Polymerization** – New nucleotides are always added only to the 3′ end of a chain. Thus, polymerization always occurs in a 5′ to 3′ direction.

a) See above.

b) The sequence of the complementary DNA would be: 3′…TAAGGC…5′. This sequence is boxed in two places on the diagram above.

c) DNA replication requires a template, and new nucleotides are added only to the 3′ end of a chain. As the two original strands are pulled apart, a template becomes available and is copied such that the new DNA is made in the 5′ to 3′ direction. At time = 2, an additional template is available and is copied such that the new DNA is made in the 5′ to 3′ direction. The directional constraints of DNA replication cause the top template strand to be copied in a discontinuous fashion.

d) See above.

(3.1) DNA replication involves many different enzymatic activities. Match each enzyme activity listed below with the function(s) that it has in the replication process. The first one is done for you.

Enzyme Activity	Function
Topoisomerase	k
Primase (synthesizes primer)	
DNA polymerase to elongate new DNA strand	
Helicase to unwind DNA	
DNA polymerase to replace RNA with DNA	
Processivity factor	

Choose from:

a) $3' \Rightarrow 5'$ growth of new DNA strand

b) $5' \Rightarrow 3'$ growth of new DNA strand

c) $3' \Rightarrow 5'$ exonuclease

d) $5' \Rightarrow 3'$ exonuclease

e) Makes RNA primer complementary to the lagging strand

f) Makes RNA primer complementary to the leading strand

g) Makes peptide bonds

h) Separates the two DNA strands

i) Maintains DNA polymerase on template

j) Provides 3'-hydroxyl for initiation of DNA polymerization

k) Untangles super-coiled DNA

(3.2) The following diagram shows a replication bubble within a DNA strand. A primer for DNA replication is 5'-GUACGUUG-3'.

a) Draw the primer where it would anneal in the replication bubble and indicate the direction of replication by an arrow.

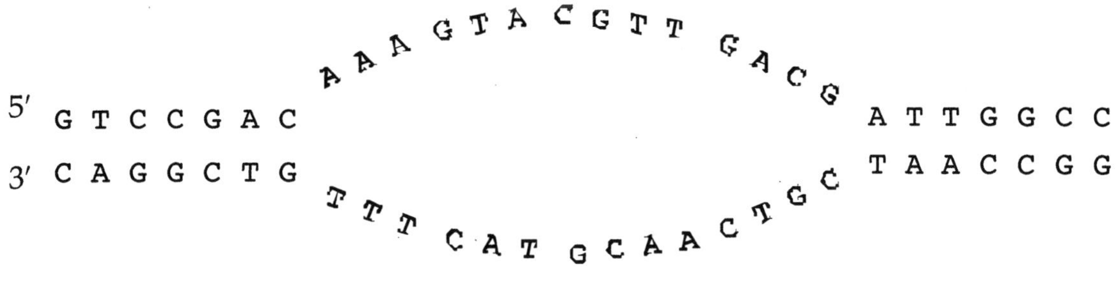

b) If the replication fork moves to the left, will this primer be used to create the leading strand or the lagging strand? Please explain your answer.

c) If you answered leading strand, explain why replication is continuous. However, if you answered lagging strand, explain why replication is discontinuous.

(3.3) Shown is a representation of an origin of replication. Synthesis of new DNA occurs on both strands and in both directions.

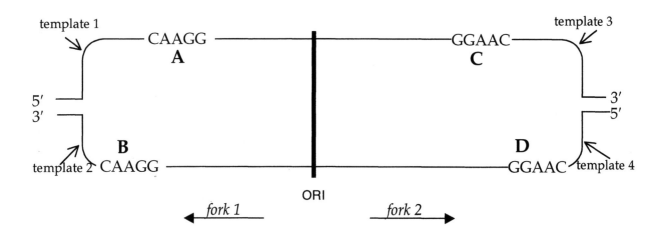

a) For the following, use sites A and B with respect to fork 1 and sites C and D with respect to fork 2.

    i) On which strand(s) will replication be continuous?

        template 1   *template 2*   *template 3*   *template 4*

    ii) To which site or sites (**A, B, C,** or **D**) can the primer 5'-GUUCC-3' bind to initiate replication?

iii) When DNA ligase is inhibited, it differentially affects the synthesis from the leading and the lagging strands. Explain which strand (leading or lagging) is more affected by the lack of DNA ligase and why.

(3.4) From this origin, replication is occurring on both strands and the two forks are moving away from each other.

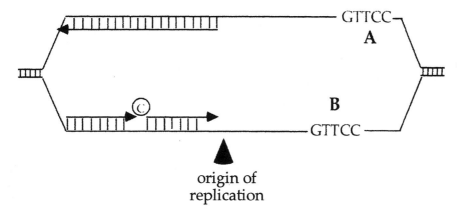

origin of
replication

a) Label the 3′ and 5′ ends of the five DNA strands shown. Indicate which strands are Okazaki fragments.

b) What enzyme is required at C in the diagram above?

c) To which site (**A** or **B** or **both**) can the primer, 5′-CAAGG-3′ bind to initiate replication?

d) For each site chosen (in iii), what is the direction of elongation (**left** or **right**) of the daughter DNA strand?

e) For each site chosen (in iii), is DNA synthesis performed in a **continuous** or a **discontinuous** fashion relative to the nearest replication fork?

(3.5) Imagine that the DNA sequence adjacent to position 90 functions as an origin of replication.

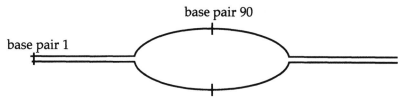

The sequence in these regions is:

```
     80                                            100
      |                                              |
5'...G C A T G C G T A C A A T A G T T C G A C ...3'
3'...C G T A C G C A T G T T A T C A A G C T G ...5'
```

a) What is the sequence of the RNA primer that binds to the top strand at base pair positions 80-90? Indicate the 5' and 3' ends of your primer.

b) Would DNA synthesis from the primer in (a) above be continuous or discontinuous?

c) What is the sequence of the RNA primer that binds to the bottom strand at base pair positions 90-100? Indicate the 5' and 3' ends of your primer.

d) Would DNA synthesis from the primer in (c) above be continuous or discontinuous?

(4) TRANSCRIPTION AND TRANSLATION

(4.1) Transcription and translation in prokaryotes

Some important notes:
- The promoter dictates where transcription starts. The promoter specifies the beginning point <u>and</u> the direction of transcription.
- The promoter is a site on the double-stranded DNA molecule where RNA polymerase binds.
- The concepts governing DNA synthesis also apply to RNA synthesis. Review directionality, base pairing, and polymerization.
- Translation starts with the AUG closest to the 5′ end of the mRNA.* This AUG need not be at the 5′ end of the mRNA, nor does it have to be a multiple of three nucleotides from the 5′ end of the mRNA.
- Translation ends with the first in-frame stop codon, even if more nucleotides remain in the mRNA.
- Translation can restart at the next start codon following a stop codon.

* Please note that this is an oversimplification used in the context of these problems. In real genes, the start codon must be preceded by a particular sequence in order to be recognized as a start codon.

Diagnostic Question:

The statement "DNA goes to RNA goes to protein" is used as a shorthand description of the flow of information in the cell.

a) "DNA goes to RNA" is meant to describe the process of _____.
For this process to occur, we need an enzyme called _____ that uses DNA as a template to make _____.

b) "RNA goes to protein" is meant to describe the process of _____.
Three different types of RNA are used in this process. _____ RNA is used as the template to make a _____. _____ RNA is part of the complex called a _____ required for protein synthesis. _____ RNA is a small RNA molecule that acts as an adapter between RNA and protein.

c) Shown below is a schematic of a prokaryotic gene. The stippled area represents the region that is transcribed.

The ▨ represents the promoter, and the direction of transcription is shown by the arrow.

a) Which strand of the DNA is being used as a template for transcription, the top or the bottom? Why?

b) On the diagram, indicate where the start codon would be. What sequence would it have?

c) On the diagram, indicate where the stop codon would be.

d) On the diagram, indicate where the transcription stop would be found.

Answer to Diagnostic Question:

The statement "DNA goes to RNA goes to protein" is used as a shorthand description of the flow of information in the cell.

a) "DNA goes to RNA" is meant to describe the process of <u>Transcription</u>. For this process to occur, we need an enzyme called <u>RNA polymerase</u> that uses DNA as a template to make <u>messenger RNA</u>.

b) "RNA goes to protein" is meant to describe the process of <u>Translation</u>. Three different types of RNA are used in this process. <u>Messenger RNA</u> is used as the template to make <u>a polypeptide (protein)</u>. <u>Ribosomal RNA</u> is part of the complex called a <u>ribosome</u> required for protein synthesis. <u>Transfer RNA</u> is a small RNA molecule that acts as an adapter between RNA and protein.

c) Shown below is a schematic of a prokaryotic gene. The stippled area represents the region that is transcribed.

The  represents the promoter, and the direction of transcription is shown by the arrow.

a) Which strand of the DNA is being used as a template for transcription, the top or the bottom?
The bottom strand is used as a template for transcription. RNA polymerase binds at the promoter and moves in the direction of the arrow. RNA can be made only in the 5' to 3' direction, antiparallel and complementary to the template. Thus, the template must be the bottom strand.

b) On the diagram, indicate where the start codon would be. What sequence would it have?
The exact position cannot be determined. In general, the start codon will be near the promoter and the sequence will be AUG.

c) On the diagram, indicate where the stop codon would be.
The exact position cannot be determined. In general, the stop codon will be near the transcription stop and could have the sequence UAA, UAG, UGA.

d) On the diagram, indicate where the transcription stop would be found.
Because you were told that the stippled area is the region transcribed, the exact position can be determined.

(4.1.1) Shown below are three genes (gene 1, gene 2, and gene 3) located on the same bacterial chromosome.

a) Indicate where on the diagram you would find the following for <u>each</u> gene:
- Promoter (p1 for gene 1, p2 for gene 2, and p3 for gene 3)
- Transcription termination site (tts1, tts2, and tts3)
- Start codon (start1, start2, and start3)
- Stop codon (stop1, stop2, and stop3)
- Template strand (ts1, ts2, and ts3), the DNA strand that directs RNA synthesis

Be sure to indicate the component on the appropriate molecule (DNA or RNA). The origin of replication is shown as a sample: the replication complex recognizes the origin of replication as a double-stranded DNA sequence (not an RNA sequence), so the origin (ori) is shown as a small region of the DNA.

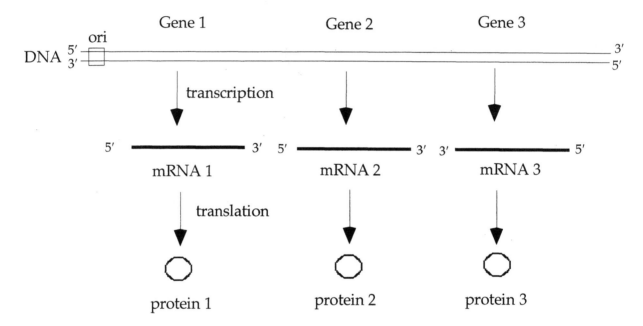

b) If a mutation inactivates the promoter for gene 2, do you expect protein 2 to be produced? Briefly explain.

c) After mutating the promoter for gene 2, you discover that protein 2 is still being made. Further analysis reveals that a spontaneous mutation has occurred in the transcription termination site in gene 1. How can you explain the presence of protein 2?

(4.1.2) Shown below is an 80 base pair segment of a hypothetical gene. It includes the promoter and the first codons of the gene. The sequences of both strands of the DNA duplex are shown: the top strand reads 5′ to 3′ left to right (1 to 80); the bottom, complimentary, strand reads 5′ to 3′ right to left (80 to 1).

a) Synthesis of the mRNA starts at the boxed A/T base pair indicated by the (a) below (#11) and proceeds left to right on the sequence below. Write the sequence of the first 10 nucleotides of the resulting mRNA.

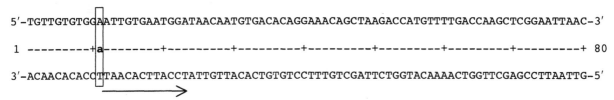

b) Suppose the synthesis of mRNA started at the boxed T/A base pair indicated by the (b) below (#77), and proceeded right to left. What would be the first five nucleotides of the mRNA?

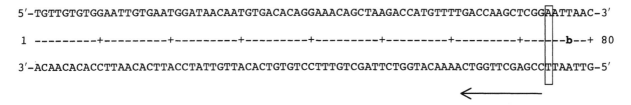

c) The mRNA you just wrote has almost the same sequence as one of the DNA strands. Which DNA strand is this? What is the difference between it and the mRNA sequence?

d) What are the first three amino acids of the polypeptide that would result from translation of the mRNA from part (a)? A table of the genetic code can be found in your textbook.

e) Does translation terminate at the UAA in the mRNA corresponding to the boxed bases at positions 48-50? Why or why not?

f) What are the last three amino acids of the polypeptide that would result from translation of the mRNA from part (a)?

g) Mutations can add base pairs, delete base pairs, or change base pairs. The sequence shown below is the same as in part (a), <u>except a G/C base pair has been added between 30 and 31</u>. Note that the overall length of the DNA is now 81 base pairs due to the addition. This is a frame-shift mutation.

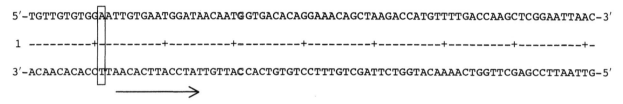

5′-TGTTGTGTGGAATTGTGAATGGATAACAATGGTGACACAGGAAACAGCTAAGACCATGTTTTGACCAAGCTCGGAATTAAC-3′

1 ---------+--------+---------+=--------+---------+---------+---------+---------+-

3′-ACAACACACCTTAACACTTACCTATTGTTACCACTGTGTCCTTTGTCGATTCTGGTACAAAACTGGTTCGAGCCTTAATTG-5′

What will the sequence of the resulting polypeptide be?

h) The sequence shown below is the same as in part (a), <u>except the C/G base pair at 27 has been deleted</u>. Note that the DNA is one base pair shorter. This is a frame-shift mutation.

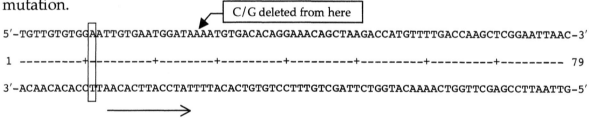

C/G deleted from here

5′-TGTTGTGTGGAATTGTGAATGGATAAAATGTGACACAGGAAACAGCTAAGACCATGTTTTGACCAAGCTCGGAATTAAC-3′

1 ---------+--------+---------+---------+---------+---------+---------+--------- 79

3′-ACAACACACCTTAACACTTACCTATTTTACACTGTGTCCTTTGTCGATTCTGGTACAAAACTGGTTCGAGCCTTAATTG-5′

What will the sequence of the resulting protein be?

i) The sequence shown below is the same as in part (a), <u>except the T/A base pair at 30 has been changed to a G/C base pair</u>.

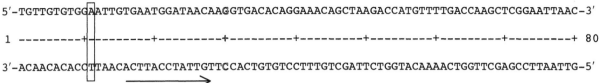

5'-TGTTGTGTGG[A]ATTGTGAATGGATAACAAGGTGACACAGGAAACAGCTAAGACCATGTTTTGACCAAGCTCGGAATTAAC-3'

1 ---------+---------+---------+---------+---------+---------+---------+---------+ 80

3'-ACAACACACC[T]TAACACTTACCTATTGTTCCACTGTGTCCTTTGTCGATTCTGGTACAAAACTGGTTCGAGCCTTAATTG-5'

What will the sequence of the resulting protein be? Is this a silent mutation (no change in amino acid sequence), missense mutation (one amino acid changed to a different amino acid), or nonsense mutation (one amino acid changed to a stop codon)?

(4.1.3) The sequences of both strands of a DNA duplex are shown below. The top strand reads 5' to 3' left to right (1 to 120) and the bottom, complementary, strand reads 5' to 3' right to left (120 to 1). The letters above or below the underlined nucleotides are keyed to the particular questions that follow.

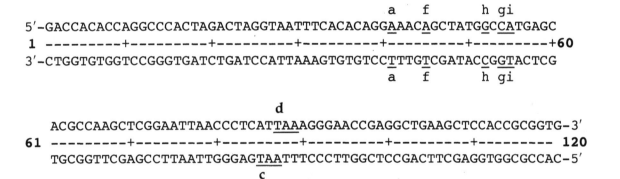

```
                                                      a     f      h gi
5'-GACCACACCAGGCCCACTAGACTAGGTAATTTCACACAGGAAACAGCTATGGCCATGAGC
1  ---------+---------+---------+---------+---------+---------+60
3'-CTGGTGTGGTCCGGGTGATCTGATCCATTAAAGTGTGTCCTTTGTCGATACCGGTACTCG
                                                      a     f      h gi
```

```
                    d
   ACGCCAAGCTCGGAATTAACCCTCATTAAAGGGAACCGAGGCTGAAGCTCCACCGCGGTG-3'
61 ---------+---------+---------+---------+---------+--------- 120
   TGCGGTTCGAGCCTTAATTGGGAGTAATTTCCCTTGGCTCCGACTTCGAGGTGGCGCCAC-5'
                    c
```

a) Assume that transcription (synthesis of mRNA) begins at the underlined A/T (written as top strand/bottom strand) at base pair 41 and proceeds to the right as shown on the diagram. What are the first 15 nucleotides of the resulting mRNA synthesized?

b) Given the mRNA synthesized in part (a), what are the first six amino acids of the resulting protein?

c) Given the mRNA synthesized in part (a), does protein translation end with the underlined TAA sequence on the bottom strand at nucleotides 85, 86, and 87? Why or why not?

d) Given the mRNA synthesized in part (a), does protein translation end with the underlined TAA sequence on the top strand at nucleotides 87, 88, and 89? Why or why not?

e) Given the mRNA synthesized in part (a), what would be the last three amino acids of the resulting protein?

f) Given the mRNA synthesized in part (a), what would the changes to the sequence of the resulting protein be if the underlined A/T base pair (written as top strand/bottom strand) at position 45 was deleted from the DNA sequence by mutation? Explain briefly.

g) Given the mRNA in part (a), what would the changes to the sequence of the resulting protein be if the underlined C/G base pair (written as top strand/bottom strand) at position 54 was changed to an A/T base pair? Explain briefly.

h) Given the mRNA synthesized in part (a), what would the changes to the sequence of the resulting protein be if the underlined G/C base pair (written as top strand/bottom strand) at position 52 was changed to a C/G base pair? Explain briefly.

i) Given the mRNA in part (a), what would the changes to the sequence of the resulting protein be if the underlined A/T base pair (written as top strand/bottom strand) at position 55 was deleted by mutation? Explain briefly.

(4.1.4) You are studying a protein and wish to determine the nucleotide sequence that encodes its first four amino acids. The sequence of the first four amino acids of the **normal** protein is as follows:

N – Met – Ser – Cys – Trp – .... – C

a) What possible mRNA sequences could encode the first four amino acids shown above? You need not write out all combinations; just write out all the codons that could encode each amino acid in their proper order.

<u>HINT</u>: there are six codons for serine (Ser). Be sure to indicate the 5' and 3' ends as appropriate.

b) You generate a mutant that produces an altered version of this protein due to a single base substitution in codon 2. The resulting protein sequence (differences from normal are **bold underlined**) is:

N – Met – **Gly** – Cys – Trp – .... – C

What does this tell you about the nucleotide sequence of the mRNA that encodes the serine (Ser) in the normal protein shown at the top of this page?

c) You generate a different mutant that produces an altered version of the protein shown at the top of this page. This mutation is due to a single base deletion of the first nucleotide of codon 2. The resulting protein sequence is shown below (differences from normal are **bold underlined**). Note that the mutation from part (b) is **not** present in this mutant.

N – Met – **Val** – **Ala** – **Gly** – .... – C

Based on this information, what is the nucleotide sequence of the mRNA that encodes the first four amino acids of the **normal** protein?

(4.1.5) You are studying a bacterium-like organism discovered on the wreck of an alien spacecraft. You discover that the creature has DNA and RNA like Earth organisms, except they have four different bases, which you name W, X, Y, and Z to differentiate them from the familiar A, T, G, and C. You also discover that the creature has transcription and translation processes almost identical to Earth cells – genes are double-stranded DNA, transcribed to mRNA, which is translated using tRNA and ribosomes.

You do notice one unusual feature: this creature's proteins are composed of only 14 amino acids, instead of the usual 20. You know that the organism's codons contain no more than three nucleotides. The amino acids used are as follows:

arginine	aspartic acid	glutamic acid	glutamine
isoleucine	leucine	lysine	methionine
phenylalanine	proline	serine	threonine
tryptophan	valine		

You decide to decipher the genetic code on this organism. First, you find conditions where you can react synthetic mRNAs with alien ribosomes, tRNA, and all the components needed to promote translation. You find that, under these unusual in vitro conditions, translation initiation can occur without a start codon and translation termination can occur without a stop codon. This means that initiation and termination can randomly occur on the mRNAs, leading to short oligopeptides.

First, you must find out how many nucleotides per codon there are.

For your first round of experiments, you add synthetic RNA to an alien translation reaction mix and sequence the protein that is synthesized. The results are as follows:

Synthetic RNA added	Protein produced
W-W-W...(W)$_n$	Met-Met-Met...(Met)$_m$
(X)$_n$	(Val)$_m$
(Y)$_n$	(Thr)$_m$
(Z)$_n$	(Leu)$_m$

a) Can you conclude anything about the number of nucleotides per codon from this information?

You perform a second round of experiments using the same procedure:

Synthetic RNA added	Protein produced
(WX)$_n$	a mixture of (Ile)$_m$ and (Glu)$_m$
(WY)$_n$	(Lys)$_m$
(WZ)$_n$	a mixture of (Arg)$_m$ and (Phe)$_m$
(XY)$_n$	a mixture of (Asp)$_m$ and (Gln)$_m$
(XZ)$_n$	a mixture of (Pro)$_m$ and (Ser)$_m$
(YZ)$_n$	(Trp)$_m$

b) From this information you are able to deduce the number of nucleotides per codon. What is this number and how did you deduce it?

You perform a final round of similar experiments:

Synthetic RNA added	Protein produced
$(WXY)_n$	$(Glu-Lys-Gln)_m$
	[note: this is equivalent to $(Gln-Glu-Lys)_m$
	and $(Lys-Gln-Glu)_m$]
$(WWX)_n$	$(Ile-Glu-Met)_m$ [same note]
$(XYZ)_n$	$(Ser-Trp-Gln)_m$ [same note]
$(WXZ)_n$	$(Pro-Glu-Arg)_m$ [same note]
$(WZY)_n$	Phe-Lys
$(XWY)_n$	Ile-Asp

Note that the last two are short peptides of only two amino acids.

c) From all the experiments above, you are able to deduce the genetic code in this organism. What is it? List all the codons and the amino acids they encode. Include stop codons.

(4.1.6) The following transcription and translation graphic is incomplete.

a) Finish the diagram by completing i to v.

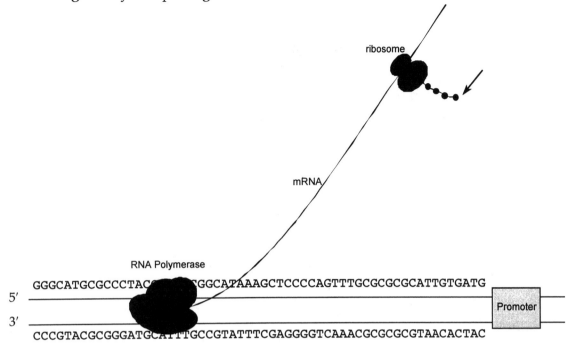

i) Label 5′ and 3′ on the mRNA.

ii) Label the arrow with either the N (amino) termini of the protein being made or the C (carboxyl) termini of the protein being made.

iii) As drawn, label the template strand for transcription.

iv) Box the three bases encoding the first amino acid of the protein being made.

v) Circle the part of the schematic where tRNAs would bind.

b) Give the anticodon used in the tRNA encoding Trp. Be sure to label the 5′ and 3′ ends.

c) Would a substitution within a codon for Trp always change the resulting protein sequence? Explain your answer.

d) Would a substitution within a codon for Thr always change the resulting protein sequence? Explain your answer.

(4.2) Transcription, RNA processing, and translation in eukaryotes

Some important notes:

The processes of DNA replication, transcription, and translation are highly conserved in prokaryotes and eukaryotes. An important difference, which is reflected in the following problems, is the structure of eukaryotic genes.

Eukaryotic genes often have DNA that is transcribed but not translated. This DNA is termed intervening sequences or introns. The DNA sequences that are represented by the mRNA that is translated into protein are considered exons.

Diagnostic Question:

a) Shown below is a schematic of the production of a polypeptide. At the top is the chromosomal arrangement found in the cell; a schematic of the polypeptide is shown below it.

   i) Label the process indicated by <u>each</u> arrow.

   ii) Indicate on the diagram below where you would expect to find each of the following components:
   - Promoter
   - Transcription terminator
   - Start codon
   - Stop codon
   - Introns
   - Exons

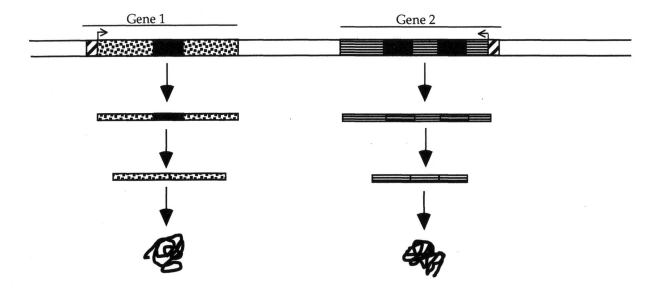

Answer to Diagnostic Question:

a) Shown below is a schematic of the production of a polypeptide. At the top is the chromosomal arrangement found in the cell; a schematic of the polypeptide is shown below it.

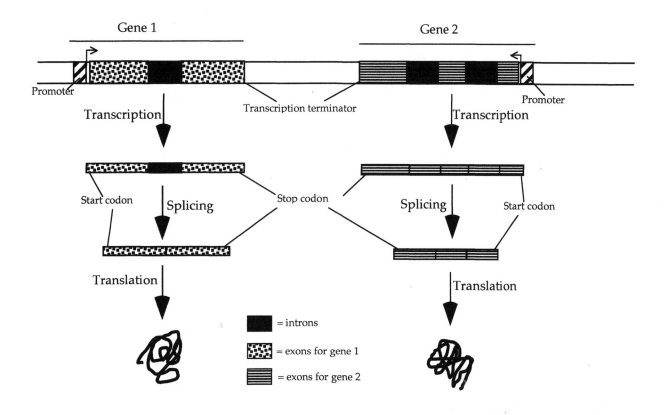

(4.2.1) Suppose that the diagram below represents the genomic organization of an enzyme involved in eye pigment production in mice. Within the gene are four exons. Biochemical analysis has revealed that the active site of the enzyme is located in the C terminus of the protein.

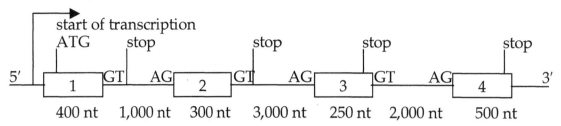

The nucleotide length of each exon and intron is shown. The dinucleotide sequence GT represents the 5′ splice site and the dinucleotide sequence AG represents the 3′ splice site. Both the 5′ and the 3′ splice sites must be present for correct splicing to occur. Assume that the first and second stop codons are located immediately after the first and second 5′ splice sites, respectively; the third and fourth stop codons are located near the 3′ end of exons 3 and 4, respectively; all these stop codons are in the correct reading frame.

a) Draw a modified version of the diagram above and indicate the nucleotide length of the mRNA transcript.

b) How many amino acids will be present in the wild-type protein produced from this transcript?

c) You are also studying the effect of various mutations in this gene.

   i) If the GT sequence (5′ splice site) located adjacent to exon 1 were deleted, what would the resulting mRNA transcript look like? Draw a similar diagram as above and indicate the nucleotide length of the mRNA transcript.

   ii) How many amino acids would be present in the protein produced from this transcript?

iii) Do you expect this enzyme to be functional? Briefly explain your answer.

d) Suppose you isolate a mutant mouse that has white eyes. When you examine the size of the eye pigment enzyme produced by this mouse, you see that it is 400 amino acids long. Sequence analysis reveals that the AG sequence (3' splice site) that is adjacent to the 5' end of exon 3 has been changed to a CC.

    i) Draw the resulting mRNA transcript that is produced as a result of this mutation.

    ii) How can you then account for the change in the size of the enzyme?

(4.2.2) The following is a picture of a eukaryotic mRNA that is base-paired with complementary single-stranded DNA.

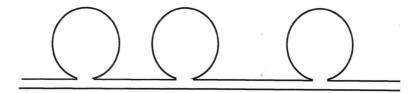

a) Label which strand is DNA and which is RNA.

b) Why does this schematic have the shape that it has?

(4.2.3) The alcohol dehydrogenase enzyme functions in the breakdown of alcohol in the liver by converting it to acetylaldehyde. The reaction is outlined below:

$$\text{Ethanol} + \text{NAD}^+ \xrightleftharpoons{\text{alcohol dehydrogenase}} \text{acetylaldehyde} + \text{NADH}$$

The enzyme binds the cofactor NAD^+ in order to carry out the oxidation of ethanol. The utilization of NAD^+ is crucial to the activity of the enzyme. The gene that encodes the 374-amino-acid enzyme is made up of 10 exons and 9 introns. Amino acid (aa) residues encoded by exons 5 to 9 are involved in binding the NAD^+ cofactor, while amino acid residues encoded by exons 1, 2, 3, 4, and 10 are involved in catalysis. A schematic diagram of the alcohol dehydrogenase gene is shown below.

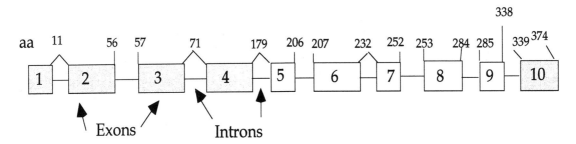

Note that codons can be interrupted by introns and later formed when the intron is excised and two exons are joined together. As an example, see amino acid 11.

a) Using a format similar to the one given in the diagram above, draw the <u>fully processed</u> mRNA of the alcohol dehydrogenase gene. Indicate the 5' and 3' ends of the mRNA.

b) The DNA structure and partial sequence of an exon/intron boundary from a region in the gene are shown below. (Note: not all 148 nucleotides in the intron are shown.)

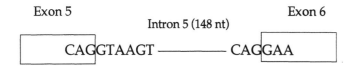

i) Using the format given above, indicate the structure and partial sequence of the mRNA at this exon/intron boundary from this region of the gene before and after splicing of the mRNA.

ii) What are amino acids 206 and 207 in the alcohol dehydrogenase enzyme?

c) Certain individuals possess a defective alcohol dehydrogenase gene. This defective gene produces a mutant enzyme. The defect is due to a mutation at the 5′ splice site in intron 5. The structure and partial sequence of this region in the DNA are shown below.

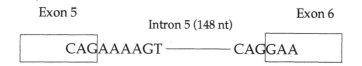

Exon 5

Intron 5 (148 nt)

Exon 6

CAGAAAAGT ——————— CAGGAA

i) Using a format similar to the one used in part (b), indicate the structure of the <u>entire</u> spliced mRNA that encodes this mutant enzyme.

ii) What two possible effects can this mutation have on the structure of the protein?

(4.2.4) The figures on the next two pages present a small gene that was found in the eukaryotic fungus *Neurospora crassa*. This is intended as a real-world example of a eukaryotic gene. Note that most genes in eukaryotes are substantially larger.

```
5'-CGGTGAATAAATACGTCATGACGGTGCTGTCAGCATCATCGATAGGTAGGAGCGAACAAACAACCTAACATCGGATTGCA
 1  +---------+---------+---------+---------+---------+---------+---------+---------
3'-GCCACTTATTTATGCAGTACTGCCACGACAGTCGTAGTAGCTATCCATCCTCGCTTGTTTGTTGGATTGTAGCCTAACGT


    GGACCGCGGGGCAGGATTGCTCCGGGCTGTTTCATGACTTGTCAGGTGGGATGACTTGGATGGAAAAGTAGAAGGTCATG
81  +---------+---------+---------+---------+---------+---------+---------+---------
    CCTGGCGCCCCGTCCTAACGAGGCCCGACAAAGTACTGAACAGTCCACCCTACTGAACCTACCTTTTCATCTTCCAGTAC


    GGGTGGCCAACTTGGGCGAGAAAAGGTATATAAAGGTCTCTTGCTCCCATCAACTGCCTCAAAAGTAGGTATTCCAGCAG
161 +---------+---------+---------+---------+---------+---------+---------+---------
    CCCACCGGTTGAACCCGCTCTTTTCCATATATTTCCAGAGAACGAGGGTAGTTGACGGAGTTTTCATCCATAAGGTCGTC


    ATCAGACAACCAAACAAACACACTTCATTCCCAAGACATCACTCACAAACAACCAACCTCTTCCAATCCAACCACAAACA
241 +---------+---------+---------+---------+---------+---------+---------+---------
    TAGTCTGTTGGTTTGTTTGTGTGAAGTAAGGGTTCTGTAGTGAGTGTTTGTTGGTTGGAGAAGGTTAGGTTGGTGTTTGT


    AAAATCAGCCAATATGTCCGACTTCGAGAACAAGAACCCCAACAACGTCCTTGGCGGACACAAGGCCACCCTTCACAACC
321 +---------+---------+---------+---------+---------+---------+---------+---------
    TTTTAGTCGGTTATACAGGCTGAAGCTCTTGTTCTTGGGGTTGTTGCAGGAACCGCCTGTGTTCCGGTGGGAAGTGTTGG


    CTAGTATGTATCCTCCTCAGAGCCTCCAGCTTCCGTCCCTCGTCGACATTTCCTTTTTTTTCATATTACATCCATCCAAG
401 +---------+---------+---------+---------+---------+---------+---------+---------
    GATCATACATAGGAGGAGTCTCGGAGGTCGAAGGCAGGGAGCAGCTGTAAAGGAAAAAAAAGTATAATGTAGGTAGGTTC


    TCCCACAATCCATGACTAACCAGAAATATCACAGATGTTTCCGAGGAAGCCAAGGAGCACTCCAAGAAGGTGCTTGAAAA
481 +---------+---------+---------+---------+---------+---------+---------+---------
    AGGGTGTTAGGTACTGATTGGTCTTTATAGTGTCTACAAAGGCTCCTTCGGTTCCTCGTGAGGTTCTTCCACGAACTTTT


    CGCCGGCGAGGCCTACGATGAGTCTTCTTCGGGCAAGACCACCACCGACGACGGCGACAAGAACCCCGGAAACGTTGCGG
561 +---------+---------+---------+---------+---------+---------+---------+---------
    GCGGCCGCTCCGGATGCTACTCAGAAGAAGCCCGTTCTGGTGGTGGCTGCTGCCGCTGTTCTTGGGGCCTTTGCAACGCC


    GAGGATACAAGGCCACCCTCAACAACCCCAAAGTGTCCGACGAGGCCAAGGAGCACGCCAAGAAGAAGCTTGACGGCCTC
641 +---------+---------+---------+---------+---------+---------+---------+---------
    CTCCTATGTTCCGGTGGGAGTTGTTGGGGTTTCACAGGCTGCTCCGGTTCCTCGTGCGGTTCTTCTTCGAACTGCCGGAG


    GAGTAAGCTCAGAGTTCACGAAAGAACCATTCGACGAGGGGAAGCACGGGGTTATCTCGTTCGAAACATGGGCCTGGTTA
721 +---------+---------+---------+---------+---------+---------+---------+---------
    CTCATTCGAGTCTCAAGTGCTTTCTTGGTAAGCTGCTCCCCTTCGTGCCCCAATAGAGCAAGCTTTGTACCCGGACCAAT


    ATGCAAATGCATAATGGGGAGGATAATGAATCATGAGGTGTACGATATGGACGATATTGACGGATCTTAATTTGATGACA
801 +---------+---------+---------+---------+---------+---------+---------+---------
    TACGTTTACGTATTACCCCTCCTATTACTTAGTACTCCACATGCTATACCTGCTATAACTGCCTAGAATTAAACTACTGT


    GTAATGAAATCACACCATAGT-3'
881 +---------+---------+  901
    CATTACTTTAGTGTGGTATCA-5'
```

Figure 1: Genomic DNA sequence of <u>con-6</u> gene from *Neurospora crassa*. The sequence of both strands (5' to 3' on top, 3' to 5' on bottom) is shown above with nucleotides numbered 1 to 901. The dashed lines are interrupted every tenth nucleotide with a "+."

```
GENOMIC DNA:    251 CAAACAAACACACTTCATTCCCAAGACATCACTCACAAACAACCAACCTC 300
                    |||||||||||||||||||||||||||||||||||||||||||||||||
       mRNA:      1 @AAACAAACACACUUCAUUCCCAAGACAUCACUCACAAACAACCAACCUC 49

GENOMIC DNA:    301 TTCCAATCCAACCACAAACAAAAATCAGCCAATATGTCCGACTTCGAGAA 350
                    |||||||||||||||||||||||||||||||||||||||||||||||||
       mRNA:     50 UUCCAAUCCAACCACAAACAAAAAUCAGCCAAUAUGUCCGACUUCGAGAA 99

GENOMIC DNA:    351 CAAGAACCCCAACAACGTCCTTGGCGGACACAAGGCCACCCTTCACAACC 400
                    |||||||||||||||||||||||||||||||||||||||||||||||||
       mRNA:    100 CAAGAACCCCAACAACGUCCUUGGCGGACACAAGGCCACCCUUCACAACC 149

GENOMIC DNA:    401 CTAGTATGTATCCTCCTCAGAGCCTCCAGCTTCCGTCCCTCGTCGACATT 450
                    |||
       mRNA:    150 CUA............................................. 152

GENOMIC DNA:    451 TCCTTTTTTTTTCATATTACATCCATCCAAGTCCCACAATCCATGACTAAC 500

       mRNA:        ................................................

GENOMIC DNA:    501 CAGAAATATCACAGATGTTTCCGAGGAAGCCAAGGAGCACTCCAAGAAGG 550
                               |||||||||||||||||||||||||||||||||||||
       mRNA:    153 ............AUGUUUCCGAGGAAGCCAAGGAGCACUCCAAGAAGG 188

GENOMIC DNA:    551 TGCTTGAAAACGCCGGCGAGGCCTACGATGAGTCTTCTTCGGGCAAGACC 600
                    |||||||||||||||||||||||||||||||||||||||||||||||||
       mRNA:    189 UGCUUGAAAACGCCGGCGAGGCCUACGAUGAGUCUUCUUCGGGCAAGACC 238

GENOMIC DNA:    601 ACCACCGACGACGGCGACAAGAACCCCGGAAACGTTGCGGGAGGATACAA 650
                    |||||||||||||||||||||||||||||||||||||||||||||||||
       mRNA:    240 ACCACCGACGACGGCGACAAGAACCCCGGAAACGUUGCGGGAGGAUACAA 288

GENOMIC DNA:    651 GGCCACCCTCAACAACCCCAAAGTGTCCGACGAGGCCAAGGAGCACGCCA 700
                    |||||||||||||||||||||||||||||||||||||||||||||||||
       mRNA:    289 GGCCACCCUCAACAACCCCAAAGUGUCCGACGAGGCCAAGGAGCACGCCA 338

GENOMIC DNA:    701 AGAAGAAGCTTGACGGCCTCGAGTAAGCTCAGAGTTCACGAAAGAACCAT 750
                    |||||||||||||||||||||||||||||||||||||||||||||||||
       mRNA:    339 AGAAGAAGCUUGACGGCCUCGAGUAAGCUCAGAGUUCACGAAAGAACCAU 388

GENOMIC DNA:    751 TCGACGAGGGGAAGCACGGGGTTATCTCGTTCGAAACATGGGCCTGGTTA 800
                    |||||||||||||||||||||||||||||||||||||||||||||||||
       mRNA:    389 UCGACGAGGGGAAGCACGGGGUUAUCUCGUUCGAAACAUGGGCCUGGUUA 438

GENOMIC DNA:    801 ATGCAAATGCATAATGGGGAGGATAATGAATCATGAGGTGTACGATATGG 850
                    |||||||||||||||||||||||||||||||||||||||||||||||||
       mRNA:    439 AUGCAAAUGCAUAAUGGGGAGGAUAAUGAAUCAUGAGGUGUACGAUAUGG 488

GENOMIC DNA:    851 ACGATATTGACGGATCTTAATTTGATGACAGTAATGAAATCACACCATAG 900
                    |||||||||||||||||||||||||||
       mRNA:    489 ACGAUAUUGACGGAUCUUAAUUUGAAAAAAAAAAAAAAAAAAAAAAAAAA 538
```

Figure 2: Sequence alignment of <u>con-6</u> genomic DNA and mRNA sequences. The top line of each pair of sequences is the sequence of <u>con-6</u> genomic DNA. The genomic DNA nucleotides are numbered as in Figure 1. The bottom line is the sequence of a <u>con-6</u> mRNA isolated from *Neurospora crassa*. The nucleotide numbers of the mRNA begin at the 5′ end with #1, and end with #539 at the 3′ end. Vertical dashes indicate nucleotides identical in both sequences (not base pairs!). Dots indicate nucleotides in the genomic sequence that are not found in the mRNA sequence (@ represents 5′ G-cap).

You should use the information to make a map of the <u>con-6</u> gene that follows the format shown below.

Maps have the following format; the numbers correspond to the numbers on the DNA strands. The map shown below is **<u>for illustration purposes only</u>** and does not correspond to the gene we are studying.

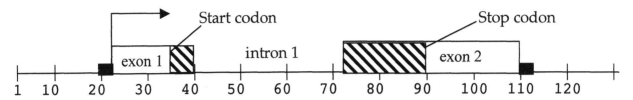

Here is what this map shows (and a list of all the features that a map must contain); note that all the map coordinates are approximate.
- The <u>promoter</u> is indicated by the small black rectangle at position 20.
- The <u>start of transcription</u> is indicated by a bent arrow; in this gene it is at position 22 (roughly).
- <u>Exon 1</u> is indicated by a labeled box; it starts at 22 and ends at 40.
- <u>Intron 1</u> is indicated by a labeled blank space; it starts at 41 and ends at 72.
- <u>Exon 2</u> is indicated by a labeled box; it starts at 73 and ends at 110.
- The <u>end of transcription</u> is indicated by the end of the last exon; here it is at 110.
- The <u>terminator</u> is indicated by a small black rectangle around position 110.
- The <u>start codon</u> is indicated by the start of the hatched region in exon 1; it is at position 35.
- The <u>stop codon</u> is indicated by the end of the hatched region in exon 2; it is at position 90.
- The <u>coding region</u>, the region that encodes the protein, is indicated by the hatched parts of the exons; it extends from 35 to 90. Note that it **<u>does not</u>** include the intron.

a) Make your map of the <u>con-6</u> gene based on Figures 1 and 2 on the preceding pages; be sure to include all the features shown on the example. If there are more than two exons and one intron, be sure to include them as well.

b) What are the first five and last five amino acids of the con-6 protein?

(4.2.5) Shown below is the sequence of a short fictitious eukaryotic gene. Both strands of DNA are shown.

After exhaustive studies, you have determined the following:

- Transcription in this organism always starts at the sixth nucleotide after the TATAA. That is, given the following sequence, the first nucleotide of the mRNA would be X: TATAAnnnnnX (where n can be any nucleotide).

- The intron splice sites have been well defined. Introns always begin with GUAUGU and end with CAG or UAG. That is, given the following sequences, the bases in **bold** will be in the mature mRNA while the underlined nucleotides will be spliced out as an intron:

XXXX<u>GUAUGU(X)ₙCAG</u>**XXXXXX**

XXXX<u>GUAUGU(X)ₙUAG</u>**XXXXXX**

- The intron-splicing machinery processes the RNA from 5′ to 3′. It finds the first GUAUGU sequence and the first subsequent (moving 5′ to 3′) CAG or UAG and removes the intron between them. (Note that these rules and sequences are similar to those of real eukaryotic organisms.)

- This organism adds poly(A) tails immediately after the sequence GAAUAAAU. Poly(A) tails are usually about seven nucleotides long.

- RNA is processed in this sequence: transcription, then splicing, then polyadenylation and capping.

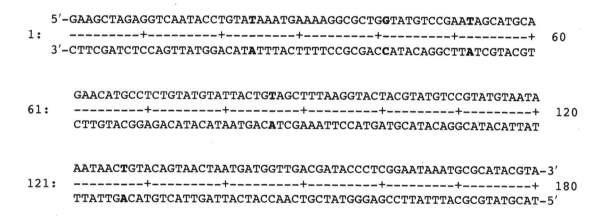

```
      5'-GAAGCTAGAGGTCAATACCTGTATAAATGAAAAGGCGCTGGTATGTCCGAATAGCATGCA
1:       ---------+---------+---------+---------+---------+---------+   60
      3'-CTTCGATCTCCAGTTATGGACATATTTACTTTTCCGCGACCATACAGGCTTATCGTACGT


         GAACATGCCTCTGTATGTATTACTGTAGCTTTAAGGTACTACGTATGTCCGTATGTAATA
61:      ---------+---------+---------+---------+---------+---------+   120
         CTTGTACGGAGACATACATAATGACATCGAAATTCCATGATGCATACAGGCATACATTAT


         AATAACTGTACAGTAACTAATGATGGTTGACGATACCCTCGGAATAAATGCGCATACGTA-3'
121:     ---------+---------+---------+---------+---------+---------+   180
         TTATTGACATGTCATTGATTACTACCAACTGCTATGGGAGCCTTATTTACGCGTATGCAT-5'
```

a) The promoter sequence is TATAA. Why wouldn't the sequence TATA (or even TATATA, for that matter) work as a promoter sequence? (Hint: remember that a promoter is not just a place for RNA polymerase to bind; a promoter must indicate which direction RNA polymerase must read the DNA.)

b) What is the sequence of the mature mRNA from this gene?

c) What is the sequence of the protein produced from this gene?

For the following mutations, describe the changes to the mRNA sequence (either list the new mRNA sequence or just list that of the altered areas) and give the sequence of the protein produced by the mutant gene. Base pairs are numbered and "C/G base pair" means: C on the top strand, G on the bottom. (The base pairs in question are highlighted in bold.) Consider each mutation separately.

d) T/A base pair 52 is changed to A/T.

e) G/C base pair 41 is changed to T/A.

f) T/A base pair 86 is changed to G/C.

g) T/A base pair 127 is changed to A/T.

h) A/T base pair 24 is deleted.

(4.2.6) This question is based on some experiments that were described by Chan, A.C. et al. in the journal *Science* (volume 264, page 1599; 1994). Shown below is an <u>internal</u> portion of the genomic DNA sequence of a gene which produces a protein kinase (we will talk about these later in the course) essential for proper immune system function, shown as double-stranded base-paired DNA:

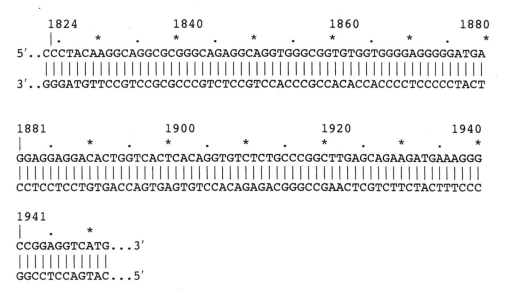

Starting with nucleotide 1824 and proceeding to the right, this region encodes the following amino acid sequence:

N-....Pro-Tyr-Lys-Lys-Met-Lys-Gly-Pro-Glu-Val-Met-...C

a) The region contains a single intron. Using the sequence data above, locate the region encoding the intron within the above genomic DNA sequence. Using the numbering scheme used above, what are the first and last nucleotides of the DNA region that encodes the intron?

b) Does the intron you have identified in part (a) follow the "GT.....AG rule" described in most textbooks (the first two nucleotides of an intron are usually GU and the last two are usually AG)? Which is more compelling evidence for the intron's boundaries, the "rule" or the protein sequence data?

c) A mutant form of this gene is known; it carries the recessive phenotype of complete immune deficiency. The mutation is a single-base-pair substitution in the region encoding the intron. The protein sequence of the corresponding region of the mutant protein is shown below with differences from wild-type underlined; the mutation results in the insertion of three amino acids into the middle of the protein.

N-...Pro-Tyr-Lys-<u>Leu-Glu-Gln</u>-Lys-Met-Lys-Gly-Pro-Glu-Val-Met-...C

i) Using the numbering scheme used above, what are the first and last nucleotides of the DNA region that encodes the intron in the mutant?

ii) Assuming that introns always end with AG and that the splicing machinery proceeds from 5' to 3', what is the mutation in these individuals (use the numbering scheme used above)?

(C2) Computer Activity 2: Gene Explorer (GeneX). This computer simulation of eukaryotic gene expression will help you to understand:
- More about transcription and translation
- Eukaryotic gene structure
- Exons and introns
- The effects of mutations on gene expression

Introduction:

Up until now, you have worked with genes of increasing complexity on paper. This can be time-consuming and error-prone. We will now move on to examine the structure and function of the genes using a computer simulation that takes care of the tedious details.

This simulation allows you to explore gene expression in a hypothetical simplified eukaryote. Eukaryotic genes have promoters and terminators for controlling transcription as well as start and stop codons for controlling translation. Although promoter and terminator sequences are different in different organisms, the genetic code, including the start and stop codons, is **identical in all organisms**. The promoter and terminator sequences used in the hypothetical organism simulated by the gene explorer are:

a promoter: a terminator:
5′–TATAA–3′   5′–GGGGG–3′
  | | | | | | | | | |
3′–ATATT–5′   3′–CCCCC–5′

Transcription begins with the first base pair to the *right* of this sequence and continues to the right.

Transcription ends with the first base pair to the *left* of this sequence.

Therefore, a gene would look like this:

5′–TATAAXXXXXXXXXXXXXXXXXXXXXXXXXXXXXXXXXXGGGGG–3′
  | | | | | | | | | | | | | | | | | | | | | | | | | | | | | | | | | | | | | | | | | | |
3′–ATATTXXXXXXXXXXXXXXXXXXXXXXXXXXXXXXXXXXCCCCC–5′

Where the region shown as X's would be transcribed into pre-mRNA.

In addition, eukaryotic genes have a few features that prokaryotic genes do not have. These are:

- Transcription produces an mRNA called a pre-mRNA that is not yet ready for translation.
- This pre-mRNA is then processed in several steps to produce the mature mRNA ready for translation:
  - The introns are removed and the exons are joined; this is called mRNA splicing. This is controlled by splice signal sequences. In real organisms, these sequences are not well known. In general, introns start with 5'-GU-3' and end with 5'-AG-3'. In the hypothetical organism simulated by the Gene Explorer, introns start with 5'-GUGCG-3' and end with 5'-CAAAG-3'.
  - A modified G nucleotide is added to the 5' end of the mRNA; this is called the "cap." In the Gene Explorer, this is not shown.
  - Many A's are added to the 3' end of the mRNA; this is called the poly(A) tail. In real organisms, as many as 400 A's can be added at a specific signal sequence; the Gene Explorer adds 13 A's as a tail to the 3' end of any mRNA. Note that these A's do not correspond to T's in the DNA.

In previous problems, you did the work of expressing a gene by hand. Now that you are familiar with how these processes work, the Gene Explorer will do all the tedious work of:

- Finding the promoter and terminator
- Reading the DNA sequence to produce the pre-mRNA
- Finding the splice sites
- Splicing and tailing the mRNA
- Finding the start codon
- Translating the mRNA

The Gene Explorer will then allow you to make specific mutations in a gene sequence, and it will then calculate and display their effects on the mRNA and protein. You do not have to deal with all the details listed above; the Gene Explorer will take care of it all. Researchers use tools like this to analyze the genes they are studying.

Procedure

Part I: The Gene Explorer

1) Start the "Gene Explorer" program in the "Molecular Biology" folder on the CD-ROM.

2) Drag the lower-right corner of the window until the Gene Explorer window fills the screen.

3) You will see something like this:

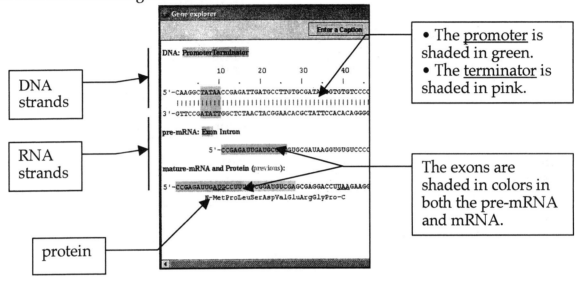

4) You can select a base in the DNA strand and the Gene Explorer will show the corresponding bases in the mRNA and the corresponding amino acid in the protein, if applicable. To select a base, here are some notes:
- You can select bases only in the top strand of DNA. If you click anywhere else on the gene panel, nothing will happen.
- It can be tricky to select a particular base on the first click; here are some tips:
  - You can use the right-arrow key to move one base to the right or the left-arrow key to move one base to the left.
  - To hit a particular base, click on the <u>space</u> between it and the next base. For example, to select 25, click on the space between 25 and 26.
- The number of the selected base is shown at the bottom right of the Gene Explorer.

An example is shown below. Here, base 63 has been selected.

This DNA base was clicked on.

This is the corresponding base in the other strand of DNA.

This is the corresponding base in the pre-mRNA; this lines up with the DNA.

This is the corresponding base in the mature mRNA. Exons are colored to match the exons in the pre-mRNA.

This is the corresponding amino acid in the protein; this lines up with the mature mRNA.

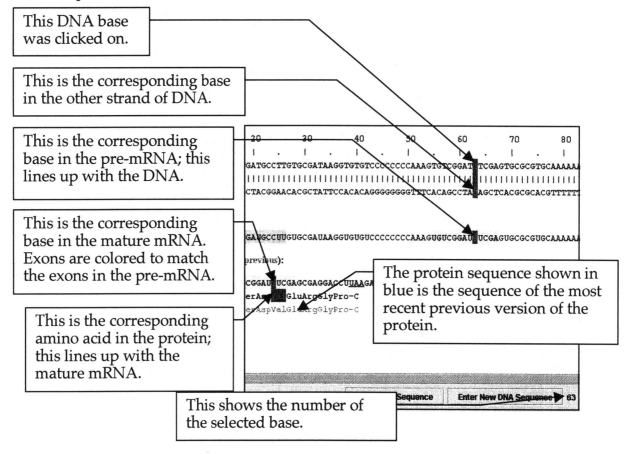

The protein sequence shown in blue is the sequence of the most recent previous version of the protein.

This shows the number of the selected base.

5) You can edit the DNA in several ways:
- To <u>delete</u> the selected base, use the "*delete*" or the "*backspace*" key.
- To <u>replace</u> the selected base with another base, type a *lowercase* letter (a, g, c, or t).
- To <u>insert</u> bases *to the left of the selected base,* type an *uppercase* letter (A, G, C, or T).

6) When you change the DNA sequence, the pre-mRNA, mature mRNA, and protein sequences are automatically updated. The *previous protein sequence*, the sequence of the protein before the latest change, is shown in blue for comparison purposes.

7) There are several useful buttons on the Gene Explorer:
- <u>Enter a Caption</u> – Click this to add a few words to the bottom of the screen. You can use this to uniquely identify your gene when you print it out from a common printer.
- <u>Print</u> – Click this to print the gene (the DNA, RNA, and protein). This prints in color and works only on Macintosh.
- <u>Print in B&W</u> – Prints in black and white. The highlighting is shown by upper- or lowercase letters. This works well on Windows and Macintosh.
- <u>Reset DNA Sequence</u> – Click this to reset the DNA to the starting gene sequence.
- <u>Enter New DNA Sequence</u> – Click this to enter a new DNA sequence.

Here is the sequence and map of the normal, starting, gene; you can use these for reference.

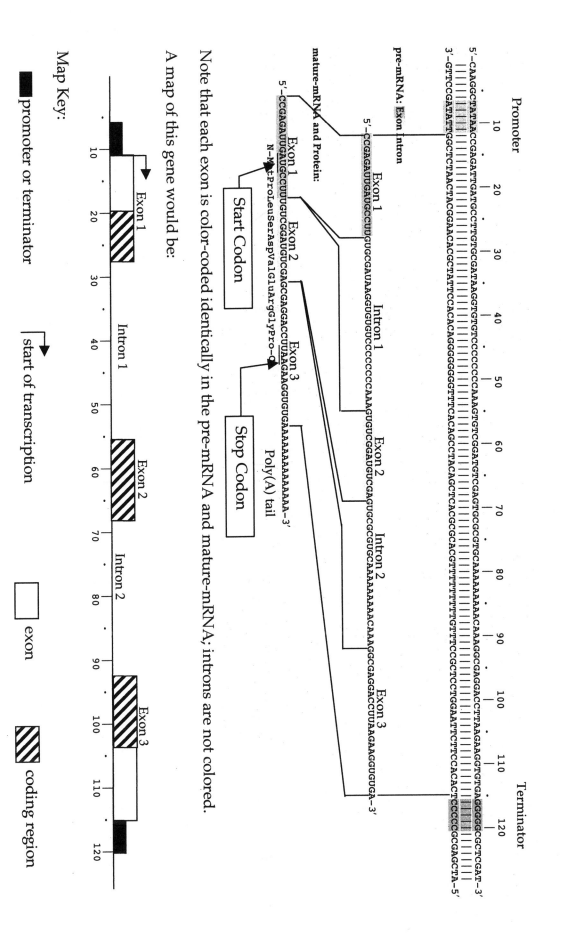

Note that each exon is color-coded identically in the pre-mRNA and mature-mRNA; introns are not colored.

A map of this gene would be:

Map Key:

▮ promoter or terminator

⌐→ start of transcription

☐ exon

▨ coding region

225

<u>Legend</u>

Gene maps like the one on the preceding page have the following format: the numbers correspond to the numbers on the DNA strands. The map shown below is **for illustration purposes only** and does not correspond to the gene we are studying.

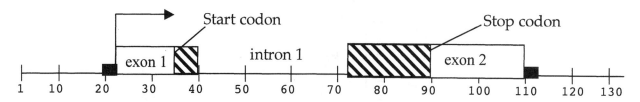

Here is what this map shows (and a list of all the features that a map must contain); note that all the map coordinates are approximate.

- The <u>promoter</u> is indicated by the small solid black rectangle at position 20.
- The <u>start of transcription</u> is indicated by a bent arrow; in this gene it is at position 22 (roughly).
- <u>Exon 1</u> is indicated by a labeled box; it starts at 22 and ends at 40.
- The <u>start codon</u> is indicated by the start of the hatched region in exon 1; it is at position 35.
- The <u>coding region</u>, the region that encodes the protein, is indicated by the hatched parts of the exons; it extends from 35 to 40 and 73 to 90. Note that it **does not** include the intron.
- <u>Intron 1</u> is indicated by a labeled blank space; it starts at 41 and ends at 72.
- <u>Exon 2</u> is indicated by a labeled box; it starts at 73 and ends at 110.
- The <u>stop codon</u> is indicated by the end of the hatched region in exon 2; it is at position 90.
- The <u>end of transcription</u> is indicated by the end of the last exon; here it is at 110.
- The <u>terminator</u> is indicated by a small solid black rectangle around position 110.

<u>Part II: A eukaryotic gene</u>

1) Look at the Gene Explorer Display; it shows the unmodified gene.

2) Click on a base in the DNA and look at the parts of the other strands that are highlighted; these correspond to the base you clicked on. You can click on any base you like or use the arrow keys to move one base to the left or right. The number of the currently selected base is shown at the bottom of the Gene Explorer window.

Use these features to explore the normal gene. For each of the following, give a DNA nucleotide number that fits the description and give the name of that part of the gene (intron, exon, untranslated region, coding region, etc.). If that type of DNA base is impossible, put impossible. There may be more than one right answer for each.

a) A DNA base that corresponds to bases in the pre-mRNA, mature mRNA, and protein.

DNA base number_____   Part of gene_____

b) A DNA base that corresponds to bases in the mature mRNA, and protein but **not** in the pre-mRNA.

DNA base number_____   Part of gene_____

c) A DNA base that corresponds to a base in the pre-mRNA, but not in the mature mRNA.

DNA base number_____   Part of gene_____

d) A DNA base that corresponds to bases in the pre-mRNA, and the mature mRNA but **not** in the protein.

DNA base number_____   Part of gene_____

3) There are several important things to notice about this gene.

a) Note that the poly(A) tail, the string of A's at the 3′ end of the mature mRNA, does not have any corresponding bases in the DNA or the pre-mRNA. How did they get there?

b) A common misconception is that introns do not split inside codons. This is not true, as you can see if you look at introns 1 and 2. Why doesn't it matter if an intron occurs in the middle of a codon?

c) Another common misconception is that an intron has to be a multiple of three nucleotides long. This is also not true, as you can see if you measure the length of introns 1 and 2 or if you try inserting some bases in an intron and see that it has no effect. Why is it that introns can be any number of nucleotides long?

Part III: Mutations

You will now make several mutations in the gene and explore their effects.

1) Click "Reset DNA Sequence."

2) Try making mutations in different parts of the normal gene; these can be insertions, deletions, or substitutions of **one DNA base.** Be sure to click "Reset DNA Sequence" before making a new mutation; that way, you can see the effects of one mutation at a time. Remember that the protein sequence corresponding to the current gene is shown in black; the previous protein sequence is shown in blue for comparison purposes.

3) Make a map of the parts of the gene that can be changed **without changing the protein sequence** (other features of the gene can change). Choose one type of mutation (insertion, deletion, or substitution) to use in your studies. Using the map below, mark off the parts of the gene that can suffer a mutation without affecting the protein sequence.

   Type of mutation: _____

4) Draw your map of regions that are insensitive to this type of mutation on the figure below:

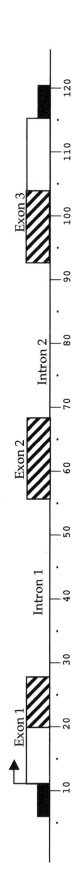

5) Why are some regions insensitive to mutation while others are sensitive to mutation?

6) Click "Reset DNA Sequence."

7) Now do mutation 1: deleting the T/A base pair at position 26.
   - Select base number 26; be sure that you have selected the right base by looking at the number in the lower right of the screen.
   - Type the delete or backspace key once.
   - Look at the sequence of the DNA strands; it should look like this:
   - If it does not look like this, go back to step 6 and try again. Note that base pair 26 is now missing.

   ```
   GCC TGTGC
       | | | | | |
   CGG ACACG
   ```

8) What is the amino acid sequence of the mutant protein? How does it differ from the original sequence? What kind of mutation is this? Remember that the previous protein sequence is shown in blue; if you follow the directions above exactly, it will show the original protein sequence.

   Original Protein Sequence: N-Met-Pro-Leu-Ser-Asp-Val-Glu-Arg-Gly-Pro-C

   Mutant Protein Sequence:

9) Using the line below, draw a map of the mutant gene. You can click on various bases in the DNA to help locate important parts of the gene. Indicate the location of the mutation in the DNA sequence with an asterisk (*). How does the structure of the mutant gene differ from the original gene?

Original Gene Map:

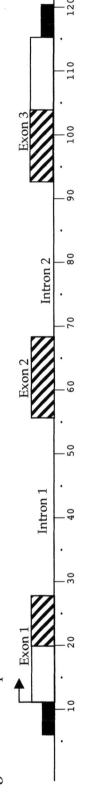

Mutant Gene Map:

10) Describe the differences between the original and mutant gene maps and explain how the mutant protein is longer even though the mutant gene is shorter.

229

11) Click "Reset DNA Sequence."

12) Now do mutation 2: changing the A/T base pair at position 51 to a T/A base pair.
   - Select base number 51; be sure that you have selected the right base by looking at the number in the lower right of the screen.
   - Type "t" (be sure it is lowercase).
   - Look at the sequence of the DNA strands; it should look like this: CCCTTAAGT / GGGAATTCA
   - Note that base pair 51 is now T in the top strand and A in the bottom strand.
   - If it does not look like this, go back to step 11 and try again.

13) What is the amino acid sequence of the mutant protein? How does it differ from the original sequence?
Original Protein Sequence: N-Met-Pro-Leu-Ser-Asp-Val-Glu-Arg-Gly-Pro-C

Mutant Protein Sequence:

14) Using the line below, draw a map of the mutant gene. Indicate the location of the mutation in the DNA sequence with an asterisk (*). How does the structure of the mutant gene differ from the original gene?

Original Gene Map:

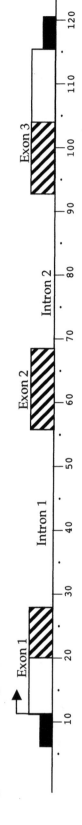

Mutant Gene Map:

15) Explain why the pattern of introns and exons is different in the mutant gene.

16) Click "Reset DNA Sequence."

17) Now do mutation 3: changing the T/A base pair at position 21 to a G/C base pair.

- Select base number 21; be sure that you have selected the right base by looking at the number in the lower right of the screen.
- Type "g" (be sure it is lowercase).
- Look at the sequence of the DNA strands; it should look like this:

```
TGAGGCCT
||| ||| |||
ACTCCGGA
```

- Note that base pair 21 is now G in the top strand and C in the bottom strand.
- If it does not look like this, go back to step 16 and try again.

18) What is the amino acid sequence of the mutant protein? How does it differ from the original sequence?

Original Protein Sequence: N-Met-Pro-Leu-Ser-Asp-Val-Glu-Arg-Gly-Pro-C

Mutant Protein Sequence:

19) Using the line below, draw a map of the mutant gene. Indicate the location of the mutation in the DNA sequence with an asterisk (*). How does the structure of the mutant gene differ from the original gene?

Original Gene Map:

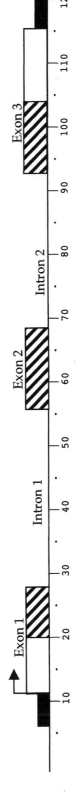

Mutant Gene Map:

20) Explain why both the start and stop codons have now moved.

231

21) Click "Reset DNA Sequence."

22) Invent a <u>nonsense mutation</u> of your own.

• Which base pair did you change? _____

• What did you change it to? _____

Original Protein Sequence: N-Met-Pro-Leu-Ser-Asp-Val-Glu-Arg-Gly-Pro-C

Mutant Protein Sequence:

• Draw a map of the resulting gene:

Original Gene Map:

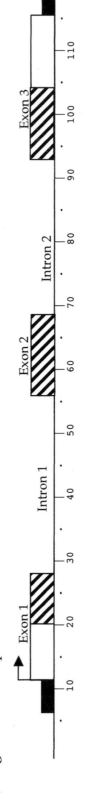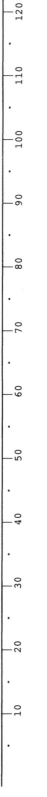

Mutant Gene Map:

23) Explain how the mutation caused the change you diagrammed in part (22).

24) Click "Reset DNA Sequence."

25) Invent a mutation where a **deletion of one base** in the DNA causes the mature mRNA to be **longer**. Note that the normal mRNA [including the poly(A) tail] is about 65 nucleotides long; you can use the tick marks on the DNA strand to estimate the length of the mature mRNA.

- Which base pair did you delete? _____

- How long is the mature mRNA in the mutant? _____

- Is the mutant protein sequence the same as the original? _____

- Draw a map of the resulting gene:

Original Gene Map:

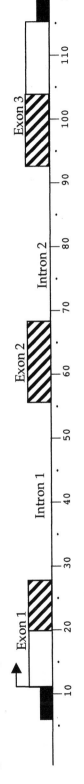

Mutant Gene Map:

26) Explain how the mutation caused the change you diagrammed in part (25).

_____

_____

27) Click "Reset DNA Sequence."

28) Invent a mutation that alters **only one DNA base** (insertion, deletion, or substitution) that results in no mRNA or protein being made at all.

- Which base pair did you change? _____

- What did you change it to? _____

- Draw a map of the resulting gene:

Original Gene Map:

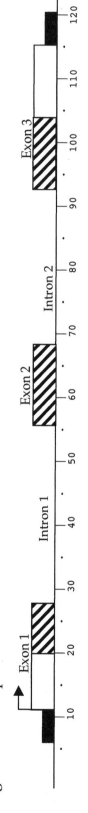

Mutant Gene Map:

29) Explain how the mutation caused no mRNA to be made and how it caused no protein to be made.

This is a very challenging problem.

30) Click "Reset DNA Sequence."

31) A single base mutation (one base inserted, deleted, or changed) in the starting gene results in the following protein sequence:

N-<u>MetProLeuSerAspVal</u>AspAlaArgAlaLysLysAsnLysGlyGluAspLeuLysLysVal-C

Note that the first six amino acids are the same as the normal protein (underlined).

What mutation led to this? (There may be more than one right answer; give only one.)

* Which base pair was changed? _____

* What was the change? _____

* Draw a map of the resulting gene:

Original Gene Map:

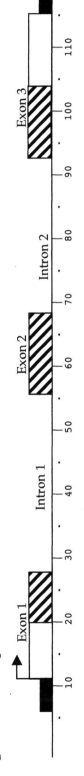

Mutant Gene Map:

32) Explain how the mutation caused the change you diagrammed in part (31).

235

33) Here are some other mutations to try. For each one, explain how the mutation has the effect described.

a) Make a mutation where one base is deleted that causes the protein to be <u>longer</u>.

b) Make a mutation where one base is inserted that causes the protein to be <u>shorter</u>.

c) Make a mutation that causes Exon 2 to be absent from the mature mRNA.

34) Design an entirely new gene that you have invented. Using the new Gene Explorer, this gene <u>should</u> (you can make it more challenging if you like):
- Produce a protein of at least five amino acids (including the N-terminal Met).
- Contain at least one intron.

<u>Tips</u>
- Use the "Enter New DNA Sequence" button and delete the starting sequence from the entry blank.
- Type in a promoter, a little DNA, and a terminator; be sure your RNA is made.
- Click on your gene and add the start codon, coding region, and stop codon; be sure your protein is made. Type slowly so that the program can keep up.
- Similarly, add an intron in the coding region and be sure your gene works.

(5) CHALLENGE PROBLEMS

(5.1) DNA synthesis cannot begin de novo. It requires a free 3'-OH group. This free OH is provided by an RNA primer. The RNA polymerase that makes this primer does not require an end on which to build. DNA polymerase's requirement for a primer has an interesting effect on DNA replication. The final 3' end of the lagging strand cannot be replicated, because there is no DNA left from which to make the RNA primer.

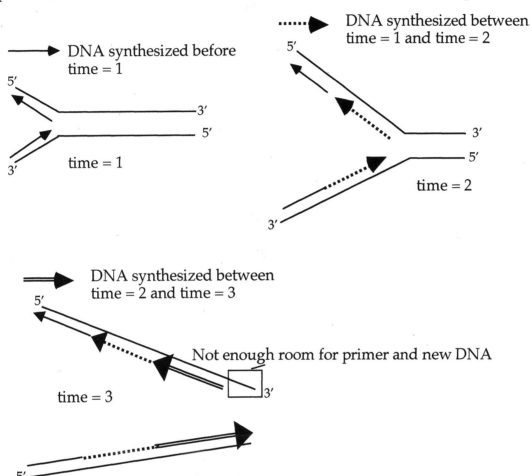

If this problem were left uncorrected, linear chromosomes (which are long stretches of DNA) would shrink with each successive replication. You might guess correctly that this would have a detrimental effect on an organism. Therefore, organisms have evolved ways to combat this problem.

a) Some organisms, such as bacteria and viruses, have circular, not linear, chromosomes. Explain how having a circular chromosome could solve the shrinkage problem explained above.

b) Another strategy that organisms take to combat shrinkage is to have linear chromosomes with telomeres at either end. A telomere is simply a long stretch of repeated nucleotides. For example, in yeast (*S. cerevisiae*), there is a telomere composed of many (TGTGTGTG)$_n$ repeats present at the end of each chromosome, where n can equal several hundred. A special enzyme called telomerase periodically extends the length of this repeat sequence without requiring a template. How could having a telomere solve the problem of shrinking chromosomes?

(5.2) Compare DNA and RNA.

a) List three key differences between DNA and RNA structures.

b) What molecular interaction allows base pairing to occur between two of the strands of DNA in a double-stranded DNA helix?

c) DNA denaturation is the separation of the two strands of a DNA molecule. Consider the two DNA sequences shown below. The symbol "|" indicates the molecular interactions between the base pairs. Which sequence (i) or (ii) would you expect to denature at a higher temperature? Briefly explain your reasoning.

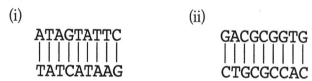

(i)
ATAGTATTC
|||||||||
TATCATAAG

(ii)
GACGCGGTG
|||||||||
CTGCGCCAC

d) A DNA helix is formed from two DNA molecules with properly aligned base pairs. Aside from base pairing, what other forces might contribute to the stability of the DNA helix?

e) The usual base pair relationships are A-T and C-G. Which bases are purines and which are pyrimidines? How might the DNA double helix be affected by the base pair mismatch of A-G or C-T.

f) In a cell that is at physiological pH, what would be the overall charge (positive or negative) of a double-stranded DNA molecule? Briefly justify your answer.

g) You carry out a DNA replication reaction using a single-stranded DNA template, DNA polymerase, a primer, and the four deoxyribonucleoside triphosphates. You then add the nucleotide form of AZT (azidothymidine) to the reaction mixture. The structure of AZT is very similar to deoxythymidine except that in AZT, the 3'-hydroxyl (OH) group on the deoxyribose ring has been replaced by an azido (N_3) group. The nucleotide form of AZT is shown below.

Nucleotide Form of AZT

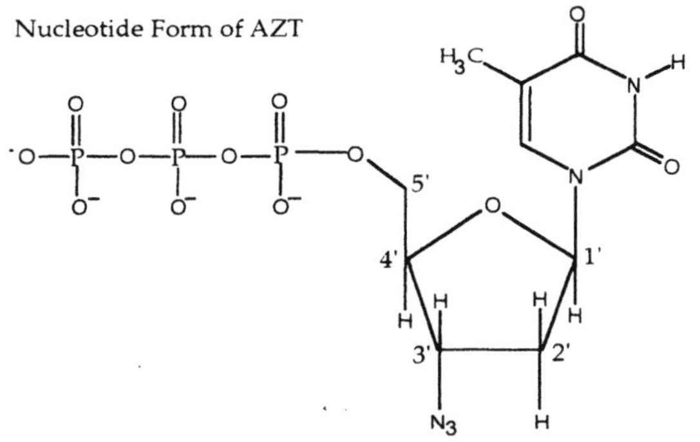

- What would you expect to happen to DNA replication when you add the AZT nucleotide to the reaction mixture? Briefly explain your reasoning.

- Why might AZT help individuals who have cancer or who are infected with HIV (human immunodeficiency virus)?

h) Consider the structure of the base, 2,6-diaminopurine, shown below.

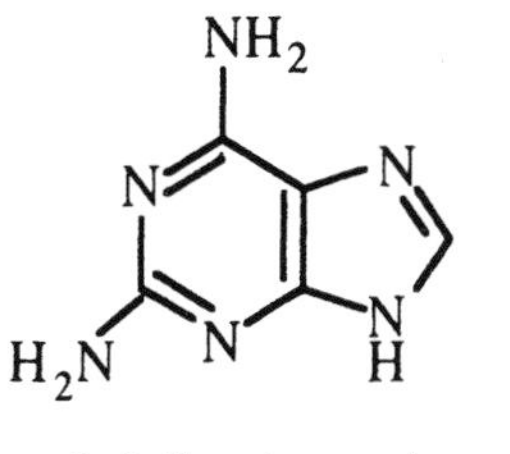

2,6-diaminopurine

With which of the normal pyrimidines (C or T) would 2,6-diaminopurine be able to base pair? Draw the base pair and indicate the hydrogen bonds that would be formed. (You do not need to draw the structure of the sugar or the phosphate groups.)

(5.3) You are studying translation in a bacterial species that has a single gene encoding the tryptophan tRNA (the *trnA-trp* gene). The wild-type sequence of this gene is shown below. The portion of the gene that encodes the anticodon of the tRNA is boxed.

```
5' GTACCTGCACTGCATGCCTAGCTAGCCCTAG CCA GCCTAGCTAGCTAGCACCAA 3'
3' CATGGACGTGACGTACGGATCGATCGGGATC GGT CGGATCGATCGATCGTGGTT 5'
```

a) Below is drawn the folded tryptophan-tRNA (produced from this *trnA-trp* gene). It is shown base pairing with an mRNA containing the codon that this tryptophan-tRNA recognizes. Fill in the three boxes on the tRNA with the correct nucleotide sequence of its anticodon. Then fill in the three boxes on the mRNA with the correct nucleotide sequence of the codon currently being read. Be sure to label the ends of the mRNA to show directionality.

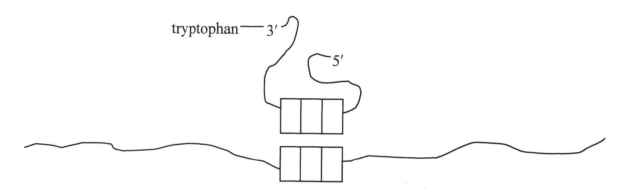

b) Which strand of the double-stranded *trnA-trp* gene is used as a template when the tryptophan tRNA is transcribed, the upper strand or the lower strand? Remember that tRNAs are transcribed directly from genes; there is no mRNA intermediate made during the production of a tRNA from its DNA sequence.

Chapter 4:

Integration Problems

In the previous chapters, we have asked you to think about biological concepts from the view of a geneticist, a biochemist, or a molecular biologist. In this chapter, we offer problems that draw from the ideas found in all three chapters. By relating ideas from these three areas, you will have the chance to practice the familiar steps in a new context. This will provide new insights into the connections between these subject areas and deepen your understanding of these important concepts.

Integration Problems:

(1) For a description of the inheritance of blood type, see your textbook and section 1.3 in the genetics section of this book.

The genes involved in production of the blood types have been studied extensively. Blood type is determined by one gene with three alleles. This gene encodes an enzyme that is involved in the synthesis of a polysaccharide on the surface of red blood cells. This enzyme is called a glycosyltransferase.

The structures of the blood type antigens (the molecules that the immune system responds to when rejecting blood of an incompatible type) are shown below:

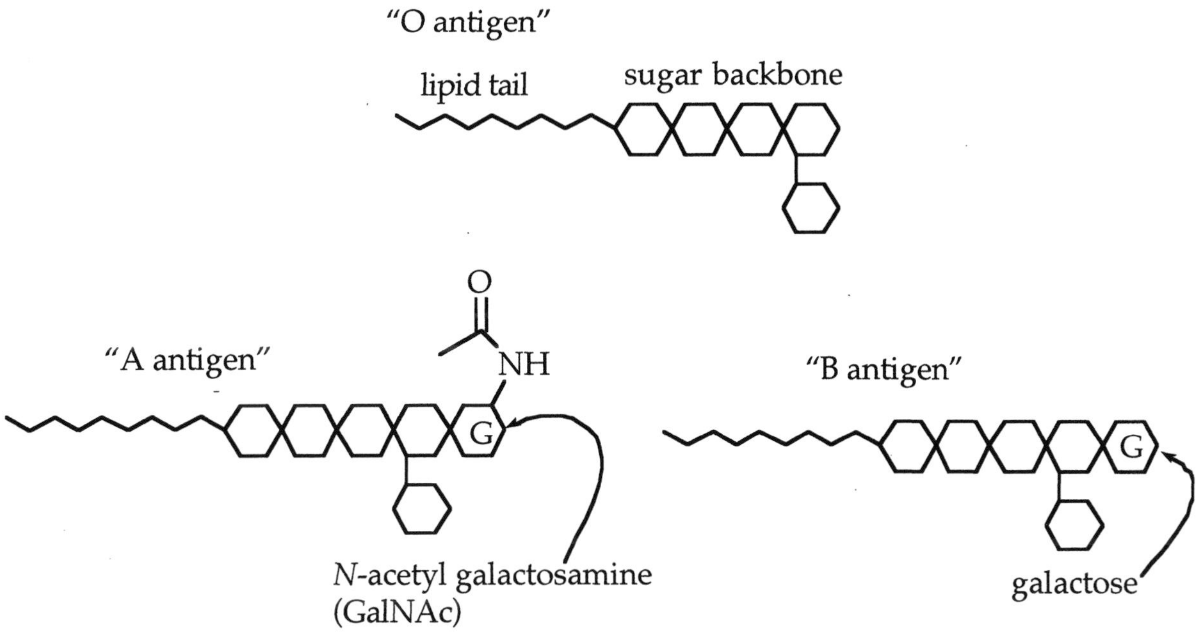

"O antigen"

lipid tail   sugar backbone

"A antigen"   "B antigen"

N-acetyl galactosamine
(GalNAc)

galactose

• The I^A allele of the blood type gene encodes a glycosyltransferase enzyme that catalyzes the following reaction:

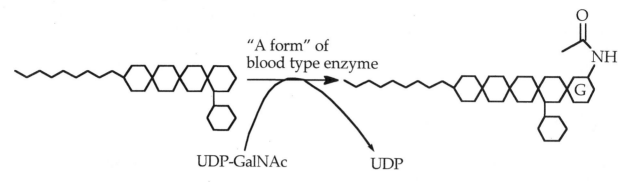

(Note: UDP is uridine diphosphate, a relative of ADP.)

• The I^B allele of the blood type gene encodes a glycosyltransferase enzyme that catalyzes the following reaction:

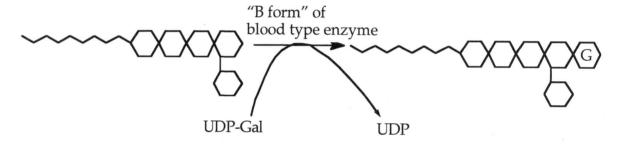

• The i allele of the blood type gene encodes a glycosyltransferase enzyme that is inactive.

a) An individual with genotype ii would not have any active glycosyltransferase. Explain in biochemical terms why an ii individual would have type O blood.

b) Explain in biochemical terms why the blood type-A phenotype of the I^A allele and the blood type-B phenotype of the I^B allele are dominant to the blood type-O phenotype of the i allele. That is, why do people with genotypes I^{A}i and I^{B}i have type A and type B blood (respectively) and not type O blood?

c) Explain in biochemical terms why the blood type-A phenotype of the I^A allele and the blood type-B phenotype of the I^B allele are codominant to each other. That is, why do people with genotype $I^A I^B$ have type AB blood (both A and B) and not A, B, or something else?

The i allele, which confers the recessive phenotype of type O blood, differs from the I^A allele by a frameshift mutation in the coding region of the gene for the blood type-determining enzyme. The DNA sequence of the coding strand (the DNA strand that has the same sequence as the mRNA, except that T's are replaced by U's) in the appropriate region of the I^A and i alleles is shown below:

> Sequence of I^A allele: CGTGGTGACCCCTT...

> Sequence of i allele: CGTGGTACCCCTT...

The relevant part of the sequence of the protein produced by the I^A and i alleles is shown below (the differences are shown in bold):

> Sequence of protein encoded by I^A allele:
>   84   85   86   87   88   89
> H₃N⁺...Leu-Val-Val-Thr-Pro-Trp-Leu...COO⁻

> Sequence of protein encoded by i allele:
> H₃N⁺...Leu-Val-Val-**Pro-Leu-Gly-Trp**...COO⁻

d) Based on the protein sequence data, indicate the reading frame of the DNA sequences above. That is, match the DNA sequences with their respective protein sequences. Note that the beginning of the reading frame must be the same in both sequences, starting from the left.

The I^A and I^B alleles differ by several point mutations, resulting in four amino acid changes in the encoded proteins. These changes are listed below:

Position in Polypeptide Chain	Amino Acid in I^A Allele	Amino Acid in I^B Allele
176	Arg	Gly
235	Gly	Ser
266	Leu	Met
268	Gly	Ala

The DNA sequence of the coding strand in the region which encodes amino acids 266 and 268 of the I^A and I^B alleles is shown below (differences shown in **bold underlined** type):

Sequence of I^A allele: ...ACTACCTGGGGGGGGTTCTT...
Sequence of I^B allele: ...ACTAC**A**TGGGGGC**C**GTTCTT...

e) Based on the mutation data, indicate the reading frame of the DNA sequences above.

f) Although the A and B glycosyltransferases differ at four places, only two of these contribute to their substrate specificity (the other two contribute only slightly to the substrate specificity). The structures of the active sites of the A and B forms of the blood type glycosyltransferase are shown below with their respective sugar substrates.

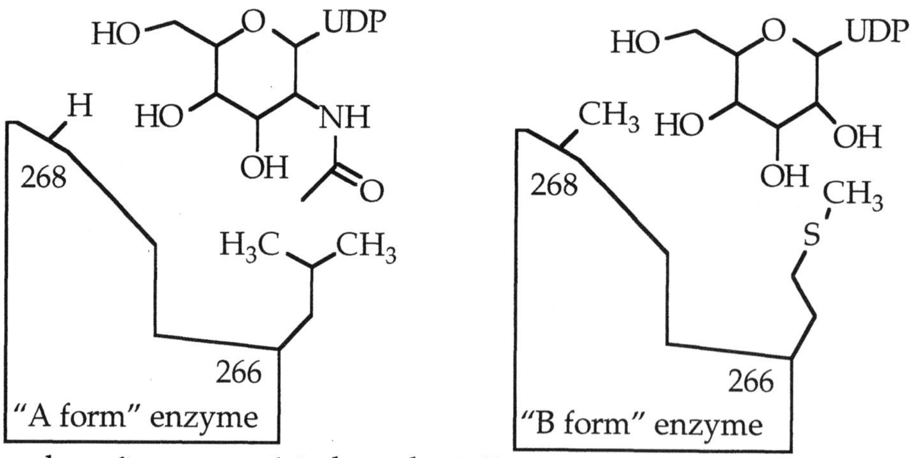

Based on these figures, explain how the different forms of the glycosyltransferase have different substrate specificities.

g) There are many rare i alleles known in the human population. For each of these mutations, provide a plausible explanation for why it would encode an inactive glycosyltransferase.

i) A mutation that changes amino acid 268 from Gly to Arg.

ii) A mutation that changes amino acid 309 from Tyr to a stop codon.

h) There is a rare allele at the blood type locus called *cis*-AB. The DNA sequence of this allele is intermediate between the I^A and I^B alleles; it contains some of the features of both. As a result, the enzyme catalyzes the reaction of the "O antigen" with **either** UDP-Gal or UDP-GalNAc. This produces the AB blood type.

i) Explain in terms of enzyme structure and function, how the changes in protein sequence in the protein encoded by the *cis*-AB allele described above could lead to the biochemical phenotype described above.

ii) Suppose that you are a researcher studying blood type. How would you tell if someone had the *cis*-AB allele (as opposed to having just the usual AB genotype $I^A I^B$)? Remember that, since you're dealing with people, you can't do crosses; you can look only at family histories (pedigrees). In other words, what blood type pedigree would indicate the presence of a *cis*-AB allele? Explain your reasoning.

(2) This problem applies genetics, biochemistry, and molecular biology to the protein hemoglobin, the protein that carries oxygen in the blood of humans. You will be given the DNA sequence of the β-globin gene from humans. This gene is located on chromosome 11. The DNA sequence includes the promoter, coding region, introns, exons, terminator, and so forth. First, you will use the protein and DNA sequences to draw a map of the major structural features of the β-globin gene. You will then be given specific mutations that have been found in this gene and will be asked to explain the effects of these mutations based on your knowledge of genetics, biochemistry, and molecular biology.

You will use the following tools as you see fit:

• Table of the genetic code. This can be found in your textbook.
• Table of amino acid structures and properties. This can also be found in your textbook.
• The program "Molecules in 3-dimensions" which you used in the Biochemistry chapter of this book to look at molecular structures in three dimensions. To launch this program, click on the panel marked "Hemoglobin," and click on the "Load Hemoglobin and show 4 chains and heme" button. The remaining buttons help you to see the amino acids relevant to the Group B mutations. In each of these views, the indicated amino acid is shown as spheres; the rest of the protein is shown as yellow dots. Consult the Biochemistry chapter of this book to find out more about how to use "Molecules in 3-dimensions."

Here is how to interpret the DNA sequence on the following pages:

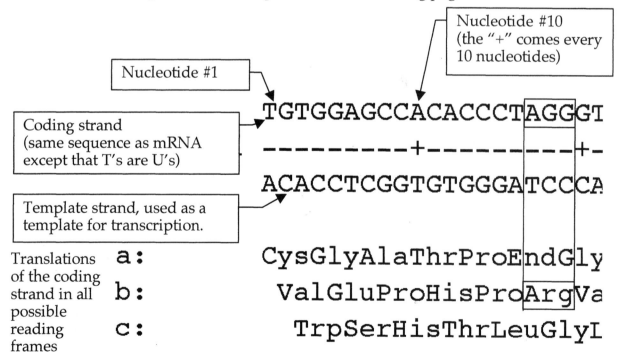

Under each line of double-stranded DNA sequence is a translation of the coding strand in all three possible reading frames. This is a convenience to save you the trouble of looking up the codons in the genetic code table. The three possible frames are:

- Frame (a) starts reading at the **<u>first</u>** nucleotide and is therefore

read as:	TGT, GGA, GCC, ACA, CCC, TAG….
or in mRNA:	UGU, GGA, GCC, ACA, CCC, UAG….
translated as:	Cys, Gly, Ala, Thr, Pro, End….

("End" = "STOP")

- Frame (b) starts reading at the **<u>second</u>** nucleotide and is therefore

read as:	GTG, GAG, CCA, …
or in mRNA:	GUG, GAG, CCA…
translated as:	Val, Glu, Pro, …

- Frame (c) starts reading at the **<u>third</u>** nucleotide and is therefore

read as:	TGG, AGC, CAC…
or in mRNA:	UGG, AGC, CAC…
translated as:	Trp, Ser, His…

Note that the sequences can be lined up in vertical columns. Therefore, if you want to find (for example) the codon and reading frame that correspond to the Arg in frame (b), just draw straight vertical lines up from each side of "Arg" to the codon as shown. This shows that the Arg was encoded by AGG preceded by a CCT in the same frame.

(I) <u>Make a map of the β-globin gene</u>

Using the amino acid sequence of the β-globin protein listed on the next page, make a map of the introns and exons in the β-globin gene. Here are some hints.

- β-globin is first made with a Met at the amino terminus (it starts from an AUG codon); that Met is removed before the protein is put into the red blood cells. The amino acids are numbered from the amino terminus of the mature protein – #1 is Val. (#0 is the starting Met.)
- The mRNA starts with nucleotide 101 and is synthesized 5' to 3' from left to right.
- The gene has three exons and two introns. Remember that introns usually start with GU and end with AG.
- The locations of the introns are marked in the protein sequence. Also, the protein sequence in the correct reading frame is underlined and the amino acids are numbered. Note that an intron can, and often does, splice in the middle of a codon.

β-globin protein sequence:

0	1	2	3	4	5	6	7	8	9	10	11	12	13	14	15	16	17
Met	Val	His	Leu	Thr	Pro	Glu	Glu	Lys	Ser	Ala	Val	Thr	Ala	Leu	Trp	Gly	Lys

Intron 1 is inserted here. ——→

18	19	20	21	22	23	24	25	26	27	28	29	30	31	32	33	34	35
Val	Asn	Val	Asp	Glu	Val	Gly	Gly	Glu	Ala	Leu	Gly	Arg	Leu	Leu	Val	Val	Tyr

36	37	38	39	40	41	42	43	44	45	46	47	48	49	50	51	52	53
Pro	Trp	Thr	Gln	Arg	Phe	Phe	Glu	Ser	Phe	Gly	Asp	Leu	Ser	Thr	Pro	Asp	Ala

54	55	56	57	58	59	60	61	62	63	64	65	66	67	68	69	70	71
Val	Met	Gly	Asn	Pro	Lys	Val	Lys	Ala	His	Gly	Lys	Lys	Val	Leu	Gly	Ala	Phe

72	73	74	75	76	77	78	79	80	81	82	83	84	85	86	87	88	89
Ser	Asp	Gly	Leu	Ala	His	Leu	Asp	Asn	Leu	Lys	Gly	Thr	Phe	Ala	Thr	Leu	Ser

Intron 2 is inserted here. ——→

90	91	92	93	94	95	96	97	98	99	100	101	102	103	104	105	106	107
Glu	Leu	His	Cys	Asp	Lys	Leu	His	Val	Asp	Pro	Glu	Asn	Phe	Arg	Leu	Leu	Gly

108	109	110	111	112	113	114	115	116	117	118	119	120	121	122	123	124	125
Asn	Val	Leu	Val	Cys	Val	Leu	Ala	His	His	Phe	Gly	Lys	Glu	Phe	Thr	Pro	Pro

126	127	128	129	130	131	132	133	134	135	136	137	138	139	140	141	142	143
Val	Gln	Ala	Ala	Tyr	Gln	Lys	Val	Val	Ala	Gly	Val	Ala	Asn	Ala	Leu	Ala	His

144	145	146
Lys	Tyr	His

a) Using the line below, draw the map of the gene encoding this portion of β-globin. Be sure to include:
• Start of mRNA
• Start and stop codons
• Introns and exons
You can use problem (4.2.4) in the Molecular Biology chapter of this book as a guide to drawing gene maps.

```
|---|---|---|---|---|---|---|---|---|---|---|---|---|---|---|---|---|
0                 500                 1000                1500
```

(II) <u>Looking at mutations</u>

There are mutations that result in the production of abnormal β-globin. Technically, the resulting disease phenotype is "β⁰ thalassemia." The "β⁰" refers to the complete absence of any β-globin protein. The precursors to red blood cells continue to make α-globin molecules. Unfortunately, in the absence of β-globin, the α-globin molecules stick together in large aggregates that destroy the red blood cells. Individuals with β⁰ thalassemia thus have no functional red blood cells and must receive frequent blood transfusions to live. β⁰ thalassemia is inherited in an autosomal recessive manner.

A list of mutations that result in β⁰ thalassemia is given below.

Mutation #	Location	DNA change	Context of change
A1	197	G ⇒ A	GT<u>G</u>GG ⇒ GT<u>A</u>GG
A2	202	A ⇒ T	GC<u>A</u>AG ⇒ GC<u>T</u>AG
A3	398	C ⇒ T	CC<u>C</u>AG ⇒ CC<u>T</u>AG
A4	170	delete A	TG<u>A</u>GG ⇒ TGGG
A5	175,176	delete AA	G<u>AA</u>GT ⇒ GGT
A6	176,177	insert G	GAAGT ⇒ GAA<u>G</u>GT

b) For each mutant,
- Give the changes in the amino acid sequence that would result from the mutation listed.

- Explain why the alteration in amino acid sequence would cause the resulting β-globin protein to be inactive.

- Explain in molecular terms why the phenotype of β⁰ thalassemia is recessive.

There are other mutations that result in hemolytic anemia. Hemolytic anemia translates as "lack of red blood cells (anemia) due to red blood cells breaking (hemolysis)." The red blood cells break open because the abnormal β-globin sticks together in large aggregates that damage the red blood cells. The phenotype of hemolytic anemia is dominant and hemolytic anemia is inherited in an autosomal manner.

A list of these mutations is given below. These change only one amino acid and have varying effects on the function of hemoglobin.

Mutation #	Location	Change	Context of change	Effect
B1	1452	G ⇒ C	TGGGC ⇒ TGCGC	hemolytic anemia
B2	233	C ⇒ A	GGCCC ⇒ GGACC	hemolytic anemia
B3	202	A ⇒ G	GCAAG ⇒ GCGAG	NORMAL HEMOGLOBIN
B4	1464	G ⇒ T	TGGTC ⇒ TGTTC	hemolytic anemia
B5	471	A ⇒ G	TCATG ⇒ TCGTG	hemolytic anemia
B6	479	A ⇒ G	AGAAA ⇒ AGGAA	hemolytic anemia

c) For each mutant,
- Give the changes in the amino acid sequence that would result from the mutation listed.

- Explain why the alteration in amino acid sequence would cause the resulting β-globin protein to be inactive.

- Explain in molecular terms why the phenotype of hemolytic anemia is dominant.

The sequence of the gene encoding β-globin follows.

DNA sequence of the β-globin gene

```
5′  TGTGGAGCCACACCCTAGGGTTGGCCAATCTACTCCCAGGAGCAGGGAGG
1   ---------+---------+---------+---------+---------+ 50
3′  ACACCTCGGTGTGGGATCCCAACCGGTTAGATGAGGGTCCTCGTCCCTCC
```

```
a:    CysGlyAlaThrProEndGlyTrpProIleTyrSerGlnGluGlnGlyGly  -
b:     ValGluProHisProArgValGlyGlnSerThrProArgSerArgGluGly-
c:      TrpSerHisThrLeuGlyLeuAlaAsnLeuLeuProGlyAlaGlyArg   -
```

```
    GCAGGAGCCAGGGCTGGGCATAAAAGTCAGGGCAGAGCCATCTATTGCTT
51  ---------+---------+---------+---------+---------+100
    CGTCCTCGGTCCCGACCCGTATTTTCAGTCCCGTCTCGGTAGATAACGAA
```

```
a:    GlnGluProGlyLeuGlyIleLysValArgAlaGluProSerIleAlaTyr-
b:     ArgSerGlnGlyTrpAlaEndLysSerGlyGlnSerHisLeuLeuLeu   -
c:    AlaGlyAlaArgAlaGlyHisLysSerGlnGlyArgAlaIleTyrCysLeu -
```

```
        ┌──→  start of mRNA
    ACATTTGCTTCTGACACAACTGTGTTCACTAGCAACCTCAAACAGACACC
101 ---------+---------+---------+---------+---------+150
    TGTAAACGAAGACTGTGTTGACACAAGTGATCGTTGGAGTTTGTCTGTGG
```

```
a:     IleCysPheEndHisAsnCysValHisEndGlnProGlnThrAspThr  -
b:    ThrPheAlaSerAspThrThrValPheThrSerAsnLeuLysGlnThrPro -
c:     HisLeuLeuLeuThrGlnLeuCysSerLeuAlaThrSerAsnArgHisHis-
```

```
    ATGGTGCACCTGACTCCTGAGGAGAAGTCTGCCGTTACTGCCCTGTGGGG
151 ---------+---------+---------+---------+---------+200
    TACCACGTGGACTGAGGACTCCTCTTCAGACGGCAATGACGGGACACCCC
    0                                   10
a:  METValHisLeuThrProGluGluLysSerALAValThrAlaLeuTrpGly -
b:   TrpCysThrEndLeuLeuArgArgSerLeuProLeuLeuProCysGlyAla-
c:    GlyAlaProAspSerEndGlyGluValCysArgTyrCysProValGly   -
```

```
    CAAGGTGAACGTGGATGAAGTTGGTGGTGAGGCCCTGGGCAGGTTGGTAT
201 ---------+---------+---------+---------+---------+250
    GTTCCACTTGCACCTACTTCAACCACCACTCCGGGACCCGTCCAACCATA
    20                                  30
a:  LysValAsnVALAspGluValGlyGlyGluAlaLeuGlyARGLeuValSer-
b:   ArgEndThrTrpMetLysLeuValValArgProTrpAlaGlyTrpTyr   -
c:    GlnGlyGluArgGlyEndSerTrpTrpEndGlyProGlyGlnValGlyIle -
```

```
        CAAGGTTACAAGACAGGTTTAAGGAGACCAATAGAAACTGGGCATGTGGA
    251 ---------+---------+---------+---------+---------+300
        GTTCCAATGTTCTGTCCAAATTCCTCTGGTTATCTTTGACCCGTACACCT

    a:      ArgLeuGlnAspArgPheLysGluThrAsnArgAsnTrpAlaCysGly  -
    b:      GlnGlyTyrLysThrGlyLeuArgArgProIleGluThrGlyHisValGlu -
    c:      LysValThrArgGlnValEndGlyAspGlnEndLysLeuGlyMetTrpArg-


        GACAGAGAAGACTCTTGGGTTTCTGATAGGCACTGACTCTCTCTGCCTAT
    301 ---------+---------+---------+---------+---------+350
        CTGTCTCTTCTGAGAACCCAAAGACTATCCGTGACTGAGAGAGACGGATA

    a:      AspArgGluAspSerTrpValSerAspArgHisEndLeuSerLeuProIle -
    b:      ThrGluLysThrLeuGlyPheLeuIleGlyThrAspSerLeuCysLeuLeu-
    c:      GlnArgArgLeuLeuGlyPheEndEndAlaLeuThrLeuSerAlaTyr  -


        TGGTCTATTTTCCCACCCTTAGGCTGCTGGTGGTCTACCTTTGGACCCAG
    351 ---------+---------+---------+---------+---------+400
        ACCAGATAAAAGGGTGGGAATCCGACGACCACCAGATGGAAACCTGGGTC
                               31
    a:      GlyLeuPheSerHisProEndAlaAlaGlyGlyLeuProLeuAspProGlu-
    b:      ValTyrPheProThrLeuArgLEULeuValValTyrProTrpThrGln  -
    c:      TrpSerIlePheProProLeuGlyCysTrpTrpSerThrProGlyProArg -


        AGGTTCTTTGAGTCCTTTGGGGATCTGTCCACTCCTGATGCTGTTATGGG
    401 ---------+---------+---------+---------+---------+450
        TCCAAGAAACTCAGGAAACCCCTAGACAGGTGAGGACTACGACAATACCC
          40                          50
    a:      ValLeuEndValLeuTrpGlySerValHisSerEndCysCysTyrGly  -
    b:      ARGPhePheGluSerPheGlyAspLeuSerTHRProAspAlaValMetGly -
    c:      GlySerLeuSerProLeuGlyIleCysProLeuLeuMetLeuLeuTrpAla-


        CAACCCTAAGGTGAAGGCTCATGGCAAGAAAGTGCTCGGTGCCTTTAGTG
    451 ---------+---------+---------+---------+---------+500
        GTTGGGATTCCACTTCCGAGTACCGTTCTTTCACGAGCCACGGAAATCAC
                    60                          70
    a:      GlnProEndGlyGluGlySerTrpGlnGluSerAlaArgCysLeuEndEnd -
    b:      AsnProLysVALLysAlaHisGlyLysLysValLeuGlyALAPheSerAsp-
    c:      ThrLeuArgEndArgLeuMetAlaArgLysCysSerValProLeuVal  -
```

254   Integration Problems

```
          ATGGCCTGGCTCACCTGGACAACCTCAAGGGCACCTTTGCCACACTGAGT
      501 ---------+---------+---------+---------+---------+550
          TACCGGACCGAGTGGACCTGTTGGAGTTCCCGTGGAAACGGTGTGACTCA
                            80
  a:      TrpProGlySerProGlyGlnProGlnGlyHisLeuCysHisThrGluEnd-
  b:       GlyLeuAlaHisLeuAspASNLeuLysGlyThrPheAlaThrLeuSer  -
  c:      MetAlaTrpLeuThrTrpThrThrSerArgAlaProLeuProHisEndVal -


          GAGCTGCACTGTGACAAGCTGCACGTGGATCCTGAGAACTTCAGGGTGAG
      551 ---------+---------+---------+---------+---------+600
          CTCGACGTGACACTGTTCGACGTGCACCTAGGACTCTTGAAGTCCCACTC
                   90                        100       104
  a:          AlaAlaLeuEndGlnAlaAlaArgGlySerEndGluLeuGlnGlyGlu  -
  b:      GLULeuHisCysAspLysLeuHisValAspPROGluAsnPheARGValSer -
  c:       SerCysThrValThrSerCysThrTrpIleLeuArgThrSerGlyEndVal-


          TCTATGGGACCCTTGATGTTTTCTTTCCCCTTCTTTTCTATGGTTAAGTT
      601 ---------+---------+---------+---------+---------+650
          AGATACCCTGGGAACTACAAAAGAAAGGGGAAGAAAGATACCAATTCAA

  a:      SerMetGlyProLeuMetPheSerPheProPhePheSerMetValLysPhe -
  b:       LeuTrpAspProEndCysPheLeuSerProSerPheLeuTrpLeuSerSer-
  c:       TyrGlyThrLeuAspValPhePheProLeuLeuPheTyrGlyEndVal  -


          CATGTCATAGGAAGGGGAGAAGTAACAGGGTACAGTTTAGAATGGGAAAC
      651 ---------+---------+---------+---------+---------+700
          GTACAGTATCCTTCCCCTCTTCATTGTCCCATGTCAAATCTTACCCTTTG

  a:       MetSerEndGluGlyGluLysEndGlnGlyThrValEndAsnGlyLysGln-
  b:        CysHisArgLysGlyArgSerAsnArgValGlnPheArgMetGlyAsn  -
  c:      HisValIleGlyArgGlyGluValThrGlyTyrSerLeuGluTrpGluThr -


          AGAUGAATGATTGCATCAGTGTGGAAGTCTCAGGATCGTTTTAGTTTCTT
      701 ---------+---------+---------+---------+---------+750
          TCTACTTACTAACGTAGTCACACCTTCAGAGTCCTAGCAAAATCAAAGAA

  a:       MetAsnAspCysIleSerValGluValSerGlySerPheEndPheLeu   -
  b:      ArgEndMetIleAlaSerValTrpLysSerGlnAspArgPheSerPhePhe -
  c:       AspGluEndLeuHisGlnCysGlySerLeuArgIleValLeuValSerPhe-
```

```
              TTATTTGCTGTTCATAACAATTGTTTTCTTTTGTTTAATTCTTGCTTTCT
         751  ---------+---------+---------+---------+---------+800
              AATAAACGACAAGTATTGTTAACAAAAGAAACAAATTAAGAACGAAAGA

         a:   LeuPheAlaValHisAsnAsnCysPheLeuLeuPheAsnSerCysPheLeu -
         b:    TyrLeuLeuPheIleThrIleValPhePheCysLeuIleLeuAlaPhePhe-
         c:     IleCysCysSerEndGlnLeuPheSerPheValEndPheLeuLeuSer   -


              TTTTTTTTCTTCTCCGCAATTTTTACTATTATACTTAATGCCTTAACATT
         801  ---------+---------+---------+---------+---------+850
              AAAAAAAAGAAGAGGCGTTAAAAATGATAATATGAATTACGGAATTGTAA

         a:    PhePheSerSerProGlnPheLeuLeuLeuTyrLeuMetProEndHisCys-
         b:     PhePheLeuLeuArgAsnPheTyrTyrTyrThrEndCysLeuAsnIle   -
         c:   PhePhePhePheSerAlaIlePheThrIleIleLeuAsnAlaLeuThrLeu -


              GTGTATAACAAAAGGAAATATCTCTGAGATACATTAAGTAACTTAAAAAA
         851  ---------+---------+---------+---------+---------+900
              CACATATTGTTTTCCTTTATAGAGACTCTATGTAATTCATTGAATTTTTT

         a:    ValEndGlnLysGluIleSerLeuArgTyrIleLysEndLeuLysLys   -
         b:   ValTyrAsnLysArgLysTyrLeuEndAspThrLeuSerAsnLeuLysLys -
         c:   CysIleThrLysGlyAsnIleSerGluIleHisEndValThrEndLysLys-


              AAACTTTACACAGTCTGCCTAGTACATTACTATTTGGAATATGTGTGTGC
         901  ---------+---------+---------+---------+---------+950
              TTTGAAATGTGTCAGACGGATCATGTAATGATAAACCTTATACACACACG

         a:   LysLeuTyrThrValCysLeuValHisTyrTyrLeuGluTyrValCysAla -
         b:    AsnPheThrGlnSerAlaEndTyrIleThrIleTrpAsnMetCysValLeu-
         c:    ThrLeuHisSerLeuProSerThrLeuLeuPheGlyIleCysValCys   -


              TTATTTGCATATTCATAATCTCCCTACTTTATTTTTCTTTTATTTTTAATT
         951  ---------+---------+---------+---------+---------+1000
              AATAAACGTATAAGTATTAGAGGGATGAAATAAAAGAAAATAAAAATTAA

         a:   TyrLeuHisIleHisAsnLeuProThrLeuPheSerPheIlePheAsnEnd-
         b:    IleCysIlePheIleIleSerLeuLeuTyrPheLeuLeuPheLeuIle   -
         c:   LeuPheAlaTyrSerEndSerProTyrPheIlePhePheTyrPheEndLeu -
```

```
             GATACATAATCATTATACATATTTATGGGTTAAAGTGTAATGTTTTAATA
      1001   ---------+---------+---------+---------+---------1050
             CTATGTATTAGTAATATGTATAAATACCCAATTTCACATTACAAAATTAT

      a:        TyrIleIleIleIleHisIleTyrGlyLeuLysCysAsnValLeuIle  -
      b:        AspThrEndSerLeuTyrIlePheMetGlyEndSerValMetPheEndTyr -
      c:        IleHisAsnHisTyrThrTyrLeuTrpValLysValEndCysPheAsnMet-


             TGTGTACACATATTGACCAAATCAGGGTAATTTTGCATTTGTAATTTTAA
      1051   ---------+---------+---------+---------+---------1100
             ACACATGTGTATAACTGGTTTAGTCCCATTAAAACGTAAACATTAAAATT

      a:        CysValHisIleLeuThrLysSerGlyEndPheCysIleCysAsnPheLys  -
      b:        ValTyrThrTyrEndProAsnGlnGlyAsnPheAlaPheValIleLeuLys-
      c:        CysThrHisIleAspGlnIleArgValIleLeuHisLeuEndPheEnd  -


             AAAATGCTTTCTTCTTTTAATATACTTTTTTGTTTATCTTATTTCTAATA
      1101   ---------+---------+---------+---------+---------1150
             TTTTACGAAAGAAGAAAATTATATGAAAAAACAAATAGAATAAAGATTAT

      a:        LysCysPheLeuLeuLeuIleTyrPhePheValTyrLeuIleSerAsnThr-
      b:        AsnAlaPhePhePheEndTyrThrPheLeuPheIleLeuPheLeuIle  -
      c:        LysMetLeuSerSerPheAsnIleLeuPheCysLeuSerTyrPheEndTyr -


             CTTTCCCTAATCTCTTTCTTTCAGGGCAATAATGATACAATGTATCATGC
      1151   ---------+---------+---------+---------+---------1200
             GAAAGGGATTAGAGAAAGAAAGTCCCGTTATTACTATGTTACATAGTACG

      a:        PheProAsnLeuPheLeuSerGlyGlnEndEndTyrAsnValSerCys  -
      b:        LeuSerLeuIleSerPhePheGlnGlyAsnAsnAspThrMetTyrHisAla -
      c:        PheProEndSerLeuSerPheArgAlaIleMetIleGlnCysIleMetPro-


             CTCTTTGCACCATTCTAAAGAATAACAGTGATAATTTCTGGGTTAAGGCA
      1201   ---------+---------+---------+---------+---------1250
             GAGAAACGTGGTAAGATTTCTTATTGTCACTATTAAAGACCCAATTCCGT

      a:        LeuPheAlaProPheEndArgIleThrValIleIleSerGlyLeuArgGln -
      b:        SerLeuHisHisSerLysGluEndGlnEndEndPheLeuGlyEndGlySer-
      c:        LeuCysThrIleLeuLysAsnAsnSerAspAsnPheTrpValLysAla  -
```

```
                GTAGCAATATTTCTGCATATAAATATTTCTGCATATAAATTGTAACTGAT
        1251    ---------+---------+---------+---------+---------1300
                CATCGTTATAAAGACGTATATTTATAAAGACGTATATTTAACATTGACTA

        a:      EndGlnTyrPheCysIleEndIlePheLeuHisIleAsnCysAsnEndCys-
        b:       SerAsnIleSerAlaTyrLysTyrPheCysIleEndIleValThrAsp  -
        c:      ValAlaIlePheLeuHisIleAsnIleSerAlaTyrLysLeuEndLeuMet -


                GTAAGAGGTTTCATATTGCTAATAGUAGCTACAATCCAGCTACCATTCTG
        1301    ---------+---------+---------+---------+---------1350
                CATTCTCCAAAGTATAACGATTATCATCGATGTTAGGTCGATGGTAAGAC

        a:      LysArgPheHisIleAlaAsnSerSerTyrAsnProAlaThrIleLeu  -
        b:      ValArgGlyPheIleLeuLeuIleValAlaThrIleGlnLeuProPheCys -
        c:      EndGluValSerTyrCysEndEndEndLeuGlnSerSerTyrHisSerAla-


                CTTTTATTTTATGGTTGGGATAAGGCTGGATTATTCTGAGTCCAAGCTAG
        1351    ---------+---------+---------+---------+---------1400
                GAAAATAAAATACCAACCCTATTCCGACCTAATAAGACTCAGGTTCGATC

        a:      LeuLeuPheTyrGlyTrpAspLysAlaGlyLeuPheEndValGlnAlaArg -
        b:       PheTyrPheMetValGlyIleArgLeuAspTyrSerGluSerLysLeuGly-
        c:       PheIleLeuTrpLeuGlyEndGlyTrpIleIleLeuSerProSerEnd  -


                GCCCTTTTGCTAATCATGTTCATACCTCTTATCTTCCTCCCACAGCTCCT
        1401    ---------+---------+---------+---------+---------1450
                CGGGAAAACGATTAGTACAAGTATGGAGAATAGAAGGAGGGTGTCGAGGA
                                                            105
        a:      ProPheCysEndSerCysSerTyrLeuLeuSerSerSerHisSerSerTrp-
        b:       ProPheAlaAsnHisValHisThrSerTyrLeuProProThrAlaPro  -
        c:      AlaLeuLeuLeuIleMetPheIleProLeuIlePheLeuProGlnLEULeu -


                GGGCAACGTGCTGGTCTGTGTGCTGGCCCATCACTTTGGCAAAGAATTCA
        1451    ---------+---------+---------+---------+---------1500
                CCCGTTGCACGACCAGACACACGACCGGGTAGTGAAACCGTTTCTTAAGT
                     110                                  120
        a:      AlaThrCysTrpSerValCysTrpProIleThrLeuAlaLysAsnSer  -
        b:      GlyGlnArgAlaGlyLeuCysAlaGlyProSerLeuTrpGlnArgIleHis -
        c:      GlyAsnValLEUValCysValLeuAlaHisHisPheGlyLYSGluPheThr-
```

```
        CCCCACCAGTGCAGGCTGCCTATCAGAAAGTGGTGGCTGGTGTGGCTAAT
1501    ---------+---------+---------+---------+---------1550
        GGGGTGGTCACGTCCGACGGATAGTCTTTCACCACCGACCACACCGATTA
                         130
a:      ProHisGlnCysArgLeuProIleArgLysTrpTrpLeuValTrpLeuMet -
b:       ProThrSerAlaGlyCysLeuSerGluSerGlyGlyTrpCysGlyEndCys-
c:        ProProValGlnAlaAlaTYRGlnLysValValAlaGlyValAlaAsn   -
          ‾‾‾‾‾‾‾‾‾‾‾‾‾‾‾‾‾‾‾‾‾‾‾‾‾‾‾‾‾‾‾‾‾‾‾‾‾‾‾‾‾‾‾‾‾‾‾‾‾‾

        GCCCTGGCCCACAAGTATCACTAAGCTCGCTTTCTTGCTGTCCAATTTCT
1551    ---------+---------+---------+---------+---------1600
        CGGGACCGGGTGTTCATAGTGATTCGAGCGAAAGAACGACAGGTTAAAGA
          140              146
a:       ProTrpProThrSerIleThrLysLeuAlaPheLeuLeuSerAsnPheTyr-
b:        ProGlyProGlnValSerLeuSerSerLeuSerCysCysProIleSer   -
c:        ALALeuAlaHisLysTyrHISEndAlaArgPheLeuAlaValGlnPheLeu -
          ‾‾‾‾‾‾‾‾‾‾‾‾‾‾‾‾‾‾‾‾‾‾‾‾‾‾‾

        ATTAAAGGTTCCTTTGTTCCCTAAGTCCAACTACTAAACTGGGGGATATT
1601    ---------+---------+---------+---------+---------1650
        TAATTTCCAAGGAAACAAGGGATTCAGGTTGATGATTTGACCCCCTATAA

a:       EndArgPheLeuCysSerLeuSerProThrThrLysLeuGlyAspIle   -
b:       IleLysGlySerPheValProEndValGlnLeuLeuAsnTrpGlyIleLeu -
c:       LeuLysValProLeuPheProLysSerAsnTyrEndThrGlyGlyTyrTyr-

        ATGAAGGGCCTTGAGCATCTGGATTCTGCCTAATAAAAAACATTTATTTT
1651    ---------+---------+---------+---------+---------1700
        TACTTCCCGGAACTCGTAGACCTAAGACGGATTATTTTTTGTAAATAAAA

a:      MetLysGlyLeuGluHisLeuAspSerAlaEndEndLysThrPheIlePhe -
b:       EndArgAlaLeuSerIleTrpIleLeuProAsnLysLysHisLeuPheSer-
c:        GluGlyProEndAlaSerGlyPheCysLeuIleLysAsnIleTyrPhe   -

        CATTGCAATGATGTATTTAAATTATTTCTGAATATTTTACTAAAAAGGGA
1701    ---------+---------+---------+---------+---------1750
        GTAACGTTACTACATAAATTTAATAAAGACTTATAAAATGATTTTTCCCT

a:       IleAlaMetMetTyrLeuAsnTyrPheEndIlePheTyrEndLysGlyAsn-
b:       LeuGlnEndCysIleEndIleIleSerGluTyrPheThrLysLysGly   -
c:       HisCysAsnAspValPheLysLeuPheLeuAsnIleLeuLeuLysArgGlu -
```

```
         ATGTGGGAGGTCAGTGCATTTAAAACATAAAGAAATGATGAGCTGTTCAA
    1751 ---------+---------+---------+---------+---------1800
         TACACCCTCCAGTCACGTAAATTTTGTATTTCTTTACTACTCGACAAGTT

a:          ValGlyGlyGlnCysIleEndAsnIleLysLysEndEndAlaValGln   -
b:          MetTrpGluValSerAlaPheLysThrEndArgAsnAspGluLeuPheLys -
c:          CysGlyArgSerValHisLeuLysHisLysGluMetMetSerCysSerAsn-



         ACCTTGGGAAAATACACTAT    3'
    1801 ---------+---------+ 1820
         TGGAACCCTTTTATGTGATA    5'

a:       ThrLeuGlyLysTyrThr      -
b:       ProTrpGluAsnThrLeu      -
c:       LeuGlyLysIleHisTyr      -
```

(3) In a fascinating and comprehensive study, Steward et al. (*Trends in Genetics* **19**[9]: 505-513 [2003]) looked at 5,686 different missense mutations, each of which led to an inheritable disease found in humans. They classified each mutation by the change in the amino acid sequence that resulted from the mutation, for example, Lys to Arg. Since there are 20 possible starting amino acids and, for each of them, there are 19 possible amino acids that they could be mutated to, there are 20 x 19 or 380 different types of missense mutations possible. They then determined how many of the 5,686 mutations fell into each category. Some types of mutation were relatively common, while others were relatively rare.

The frequency with which a given type of mutation leads to disease depends on two factors:
- How likely it is that a mutation could lead to that change
- How damaging that mutational change would be to the protein

The chance that a random mutation could lead to a particular change depends on the genetic code; for example, changing Gly (GGG) to Arg (AGG) requires only one base to be mutated, while Phe (UUU) to Asn (AAU) requires two bases to be mutated. Since changing two bases is much less likely than changing one, Gly to Arg mutations will occur more often than Phe to Asn mutations.

In fact, mutations are not completely random. In humans, it turns out that certain mutations occur more frequently than others. In humans, the C bases in CG sequences are sometimes modified by the addition of a methyl group. At a low frequency, these methyl-C's undergo spontaneous deamination and become T's. If this is not properly repaired, the GC sequence can become a CA or a TG sequence depending on which strand the methyl-C was in. Other mutations are known to occur at slightly lower frequencies.

The degree of damage that a particular amino acid change would do to a given protein depends on the properties of different amino acid side chains and their interactions as they influence protein structure and function. Remember that for a mutation to result in a genetic disease, it must have a substantial (usually negative) effect on the protein's function.

Using these factors, explain the following observations.

a) The most common type of mutation (229 of the 5,686 total) is Arg to Cys. Why would you expect this to be so frequent?

b) Another frequent type is Arg to Trp (197 of 5,686). Why would you expect this to be so frequent?

c) Another frequent type is Arg to His (217 of 5,686). Why is it surprising that this is so frequent?

d) The mutation Val to Pro was never observed in their set of 5,686 mutations. Why would you expect that it would be very infrequent?

e) The mutation Leu to Ile was very infrequent (3 of 5,686). Why would you expect that this would be rare?

f) The mutation Gly to Phe was never observed in their set of 5,686 mutations. Why would you expect that this mutation would be very infrequent?

(4) Below is the DNA sequence of the first part of a hypothetical gene. The promoter is underlined and transcription begins at and includes the bold G/C base pair.

```
5' TACAC GCTTA GCTGA CTATA AGGAC GAATC GCTAC AACGA TGCGA-
   ||||| ||||| ||||| ||||| ||||| ||||| ||||| ||||| |||||
3' ATGTG CGAAT CGACT GATAT TCCTG CTTAG CGATG TTGCT ACGCT-


   -TGCCA TCCGA TTGGT GTTCC TTCCA TGAAG GATGC ACAAC GCAAA 3'
    ||||| ||||| ||||| ||||| ||||| ||||| ||||| ||||| |||||
   -ACGGT AGGCT AACCA CAAGG AAGGT ACTTC CTACG TGTTG CGTTT 5'
```

a) What are the first 12 nucleotides of the transcript encoded by this gene? Label the 5' and 3' ends.

b) On the DNA sequence above, **circle** the DNA bases that encode the first amino acid of the protein.

c) What are the first four amino acids encoded by this gene? Label the N and C termini.

d) You want to create a system to translate a specific mRNA in a test tube. To an appropriate water and salt solution you add many copies of this mRNA and ATP. What other key components must you add?

You succeed in translating the mRNA in your test tube. You repeat the experiment with two identical test tubes. You add trace amounts of the antibiotic puromycin to test tube 2 only. Puromycin is a molecule that has structural similarities to the 3' end of a charged tRNA. It can enter the ribosome and be incorporated into the growing protein. When puromycin is incorporated into the polypeptide, it stalls the ribosome and the polypeptide is released.

e) What effect would puromycin have on transcription?

f) What effect would puromycin have on translation?

g) In principle, there are two possible modes through which puromycin could bind to the ribosome:
  A) Puromycin binds to a specific codon in the mRNA and stops translation there.
  B) Puromycin does not bind to a specific codon and stops translation wherever the ribosome happens to be when the puromycin binds.

To distinguish between these hypotheses, you set up two test tubes:
  • Test tube 1: The same translation mixture you described above.
  • Test tube 2: The translation mixture with puromycin added.

You allow translation to occur for a little while and then examine the length of the polypeptide produced in both test tubes.

i) In test tube 1 you get a polypeptide that is 100 amino acids long. At least how many bases long was the complete mRNA that you added?

ii) Suppose that model (A) is correct. What would you expect to find in test tube 2?
  • Only a single type of polypeptide.
  • Only two types of polypeptides that are each different lengths.
  • Only three types of polypeptides that are each different lengths.
  • Only four types of polypeptides that are each different lengths.
  • Polypeptides of all sizes, that is, dipeptides, tripeptides, … a polypeptide that is 100 amino acids long.
Explain your reasoning.

iii) Suppose that model (B) is correct. What would you expect to find in test tube 2?

- Only a single type of polypeptide.
- Only two types of polypeptides that are each different lengths.
- Only three types of polypeptides that are each different lengths.
- Only four types of polypeptides that are each different lengths.
- Polypeptides of all sizes, that is, dipeptides, tripeptides, … a polypeptide that is 100 amino acids long.

Explain your reasoning.

This gene encodes a protein that binds to the neurotransmitter serotonin, as shown below. The five amino acids involved in binding serotonin are shown.

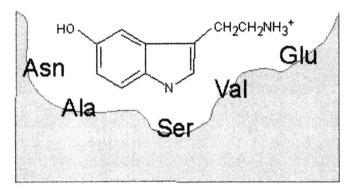

Below is an <u>internal</u> part of the wild-type DNA sequence and the protein it encodes. The amino acids depicted in the picture above are <u>underlined</u>.

```
DNA        5'...ACC AAT GGA CCA GCA GGA AGC GGG GTA GCT GAG TAC...3'
               ||| ||| ||| ||| ||| ||| ||| ||| ||| ||| ||| |||
           3'...TGG TTA CCT GGT CGT CCT TCG CCC CAT CGA CTC ATG...5'

Protein    N-...Thr Asn Gly Pro Ala Gly Ser Gly Val Ala Glu Tyr...—C
```

h) You find the following alternative DNA sequence for this protein:

```
5'...ACC AAT GGA CCA GCA GGA TAG CGG GGT AGC TGA GTAC...3'
       ||| ||| ||| ||| ||| ||| ||| ||| ||| ||| ||| ||||
3'...TGG TTA CCT GGT CGT CCT ATC GCC CCA TCG ACT CATG...5'
```

i) Indicate (circle/underline) the site of the mutation on the sequence directly above.

ii) Does the alternative sequence have an insertion, deletion, or substitution mutation?

iii) Would you expect this DNA sequence to encode a protein that binds serotonin? Why or why not?

i) You find a third DNA sequence for this protein

```
5'...ACC AAT GGA CCA GCA GGA AGC GGG GTA GCT GAT TAC...3'
       ||| ||| ||| ||| ||| ||| ||| ||| ||| ||| ||| |||
3'...TGG TTA CCT GGT CGT CCT TCG CCC CAT CGA CTA ATG...5'
```

i) Indicate (circle/underline) the site of the mutation on the above sequence.

ii) Does this third sequence have an insertion, deletion, or substitution mutation?

iii) Would you expect this DNA sequence to encode a protein that binds serotonin? Why or why not?